An experimental and numerical study of low salinity effects on the oil recovery of carbonate limestone samples

Clausthal University of Technology

AN EXPERIMENTAL AND NUMERICAL STUDY OF LOW SALINITY EFFECTS ON THE OIL RECOVERY OF CARBONATE LIMESTONE SAMPLES

Doctoral Thesis
(Dissertation)

to be Awarded the Degree
Doctor of Engineering (Dr.-Ing.)

submitted by
M.Sc. Felix Feldmann
from Oelde, Germany

Approved by the Faculty of
Energy and Economic Sciences,
Clausthal University of Technology

Date of oral Examination
January 30, 2020

Bibliografische Information der Deutschen Nationalbibliothek
Die Deutsche Nationalbibliothek verzeichnet diese Publikation in der
Deutschen Nationalbibliografie; detaillierte bibliografische Daten sind
im Internet über http://dnb.d-nb.de abrufbar.
1. Aufl. - Göttingen: Cuvillier, 2020
Zugl.: (TU) Clausthal, Univ., Diss., 2020

Dissertation Technische Universität Clausthal - 2020

Dean of the Faculty of Energy and Economic Sciences:
Prof. Dr. rer. nat. Bernd Lehmann

Chairperson of the Board of Examiners:
Prof. Dr. rer. nat. Bernd Lehmann

Supervising tutor:
Prof. Dr. mont. Leonhard Ganzer

External reviewers:
Prof. Dr. rer. nat. Holger Ott
Prof. Dr. mont. Martin Bremeier
Dr. Ali. M. AlSumaiti

© CUVILLIER VERLAG, Göttingen 2020
Nonnenstieg 8, 37075 Göttingen
Telefon: 0551-54724-0
Telefax: 0551-54724-21
www.cuvillier.de

Alle Rechte vorbehalten. Ohne ausdrückliche Genehmigung
des Verlages ist es nicht gestattet, das Buch oder Teile
daraus auf fotomechanischem Weg (Fotokopie, Mikrokopie)
zu vervielfältigen.
1. Auflage, 2020
Gedruckt auf umweltfreundlichem, säurefreiem Papier
aus nachhaltiger Forstwirtschaft

ISBN 978-3-7369-7176-9
eISBN 978-3-7369-6176-0

Abstract

Low-salinity waterflooding is a relatively simple and cheap Enhanced oil recovery technique in which the salinity of the injected water is optimized (by desalination and/or modification) to improve oil recovery over conventional waterflooding. Extensive experimental studies that are investigating the effect of low-salinity waterfloodings are available in the literature. Sulfate-rich as well as diluted brines have shown promising potential to increase oil production in limestone core samples. To quantify the low-salinity effect, spontaneous imbibition and/or waterflooding experiments have been reported. This thesis combines spontaneous imbibition, centrifuge method and unsteady state coreflooding experiments to investigate low-salinity effects in carbonate samples. Moreover, zeta potential measurements were conducted to evaluate the concept of Surface charge change as the driving low-salinity mechanism in carbonates.

The experimental study uses three brine compositions to investigate possible low-salinity effects: A high-saline Formation-water (salinity of $183.4\,g/l$), Sea-water (salinity of $43.8\,g/l$) and 100 times Diluted-sea-water (salinity of $0.44\,g/l$). Initially, a sequence of spontaneous imbibition experiments was conducted to screen the impact of connate and imbibing water composition on spontaneous oil recovery. After completing the spontaneous imbibition tests, the samples were drained inside a centrifuge to determine the impact of brine composition on residual saturation and capillary pressure. The study was completed by the conduction of three unsteady state corefloodings to simulate the water-oil displacement under more field-realistic conditions. Thereby, each brine was tested in secondary mode (brine injection at connate water saturation) and tertiary mode application (exchange of injection brine at a mature recovery stage).

Spontaneous imbibition experiments only showed oil recovery in case the salinity of the imbibing water was lower than the salinity of the connate water. No oil production was observed when the imbibing water had a higher salinity than the connate water, or the salinity of the connate water and imbibing brine were identical. Using Formation-water as connate water, Diluted-sea-water resulted in 35.4 % spontaneous oil recovery, Sea-water in 22.1 % spontaneous oil recovery and Formation-water in 1.5 % spontaneous oil recovery.

The centrifuge experiments confirmed a correlation between the overall system's salinity and oil recovery. As the salinity of the imbibing brine decreased, the capillary imbibition pressure curve showed an increasing water-wetting tendency and simultaneous reduction of the remaining oil saturation. In the case of Formation-water as connate water, the forced imbibition resulted in a residual oil saturation of 5.6 % for Diluted-sea-water, 10.7 % for Sea-water and 15.3 % for Formation-water. The lowest remaining oil saturation was obtained for Diluted-sea-water as connate water and imbibing water.

The coreflooding experiments reflected the results of the spontaneous imbibition and centrifuge experiments. Comparing the oil production at field rate equivalent injection of $0.05\,cm^3/min$, Sea-water and especially Diluted-sea-water resulted in a significantly higher oil recovery than Formation-water. Moreover, when comparing secondary mode experiments, the injection of Diluted-sea-water resulted in a remaining oil saturation of 30.6 %, Sea-water in a remaining oil saturation of 35.5 % and Formation-water in a remaining oil saturation of 37.4 %. In tertiary injection mode, Sea-water did not lead into extra oil recovery while Diluted-sea-water caused an additional oil recovery of 5.6 % in one out of two tertiary injection applications.

The zeta potential measurement supported the concept of Surface charge change as the driving mechanisms of low-salinity effects in limestones. Compared to Formation-water and Sea-water, the Diluted-sea-water and limestone systems exhibited the strongest negative zeta potential. In line with the literature review and the results of the spontaneous imbibition, centrifuge method and corefloodings, Surface charge change is a reasonable explanation of wettability alteration in limestones.

To validate and history match the experimental obtained data, a numerical centrifuge and coreflooding model was implemented inside the C++ DuMu^x simulator. The mathematical model descriptions include the imposed boundary conditions, fluid properties, hydraulic properties, flow model and material balance coupling. The presented numerical centrifuge and coreflooding models were validated against the commercial Cydar simulator.

It is shown that the numerical centrifuge simulation confirms the analytical imbibition capillary data analysis. An increasing water-wetting tendency and a simultaneous reduction of the residual oil saturation was observed, in case the salinity of the imbibition brines decreased in comparison to Formation-water.

In line with the centrifuge results, the numerical coreflooding simulations confirmed a correlation between salinity, wettability and oil recovery. In secondary injection mode, the numerically obtained relative permeability showed the strongest water-wet behavior for Diluted-sea-water, followed by Sea-water and Formation-water.

Kurzfassung

Low-salinity Wasserfluten ist eine simple und kostengünstige Technik zur verbesserten Ölgewinnung, bei der der Salzgehalt des eingepressten Wassers (durch Entsalzung und/ oder Modifikation) optimiert wird. In der Fachliteratur sind experimentelle Studien verfügbar, in denen sulfatreiches als auch entsalztes Wasser ein vielversprechendes Potenzial zur Steigerung der Ölproduktion aus Kalksteinproben gezeigt haben. Diese Dissertation kombiniert erstmalig Spontaneous-imbibition, Zentrifugen und Kernflutungsexperimente zur Untersuchung von Low-salinity Effekten in Kalksteinproben. Darüber hinaus wurden Zeta-Potential Messungen durchgeführt, um das Konzept der Oberflächenladungsveränderung als möglichen Mechanismus von Low-salinity Effekten in Kalksteinen zu untersuchen.

In der experimentellen Studie wurden drei Injektionswasserzusammensetzungen getestet: Ein hochsalzhaltiges Formationswasser (Salzgehalt von 183,4 g/l), ein sulfatreiches Meerwasser (Salzgehalt von 43,8 g/l) und 100-fach verdünntes Meerwasser (Salzgehalt von 0,44 g/l). Die Spontaneous-imbibition Tests zeigten nur dann Ölproduktion, wenn der Salzgehalt des eindringenden Wassers niedriger war als der Salzgehalt des initialen Wassers. Es wurde keine Ölproduktion beobachtet, wenn das eindringende Wasser einen höheren Salzgehalt als das initiale Wasser aufwies oder der Salzgehalt der beiden Fluide identisch war. Die Spontaneous-imbibition Tests ergaben eine Ölgewinnung von 35,4 % für 100-fach verdünntes Meerwasser, 22,1 % für Meerwasser und 1,5 % für Formationswasser.

Die Zentrifugenversuche bestätigten eine Verbindung zwischen Salzgehalt und Ölgewinnung. Mit abnehmendem Salzgehalt zeigten die Kapillardruckkurven eine zunehmende Wasserbenetzungstendenz und eine gleichzeitige Verringerung der Rest- ölsättigung. Die gemessenen Restölsättigungen waren 5,6 % für verdünntes Meerwasser, 10,7 % für Meerwasser und 15,3 % für Formationswasser.

Die Kernflutungen spiegelten die Ergebnisse der Spontaneous-imbition Tests und der Zentrifugenversuche wider. Unter lagerstättenrealistischer Injektionsrate resultierte Meerwasser und insbesondere verdünntes Meerwasser in eine signifikant höhere Ölproduktion als Formationswasser. Nach dem Erhöhen der Injektionsrate ergab verdünntes Meerwasser eine verbleibende Ölsättigung von 30,6 %, Meerwasser von 35,5 % und Formationswasser von 37,4 %.

Die durchgeführten Zeta-Potential Messungen stützen das Konzept der Oberflächenladungsveränderung als verantwortlichem Mechanismus von Low-salinity Effekten in Kalksteinen. Im Vergleich zu Formationswasser und Meerwasser zeigte das System aus verdünntem Meerwasser und Kalkstein die stärkste negative Oberflächenladung.

Im Bezug auf die Literaturrecherche, den Spontaneousimbition, Zentrifugen und Kernflutungsexperimenten kann geschlussfolgert werden, dass die Injektion von verdünntem Meerwasser die Benetzung von Kalksteinen verändert.

Um die experimentellen Daten zu überprüfen, wurde ein numerisches Zentrifugen- und Kernflutungsmodell im C++ Simulator Dumux implementiert. Die mathematischen Modellbeschreibungen umfassen die definierten Randbedingungen, hydraulischen Eigenschaften, das Strömungsmodell und die Materialbilanzkopplung. Die entwickelten Modelle wurden mit Hilfe einer kommerziellen Simulationssoftware validiert.

Die numerischen Modelle bestätigten die analytische Zentrifugenauswertung. Die Kapillardruckkurven sind durch eine zunehmende Wasserbenetzungstendenz und eine gleichzeitige Verringerung der Restölsättigung gekennzeichnet, wenn der Salzgehalt des Injektionswasser im Vergleich zu Formationswasser abnimmt.

Die Kernflutungssimulationen bestätigen eine Verbindung von Salzgehalt, Benetzbarkeit und Ölgewinnung. Die numerisch abgleiten relative Permeabilitatën zeigten die stärkste Wasserbenetzung für verdünntes Meerwasser, gefolgt von Meerwasser und Formationswasser.

Acknowledgements

Foremost, I would like to express my sincere gratitude to Dr. Ali M. AlSumaiti for the continuous support of my research and Doctoral studies. For his patience, motivation, enthusiasm, trust and knowledge. His guidance helped me during all the time of research and writing of this dissertation.

I would especially like to thank Professor Dr. Leonhard Ganzer for his guidance and support throughout my Bachelor, Master and Doctoral studies. For sending me abroad, always being optimistic and believing in the progress of my work.

Professor Dr. Holger Ott for reviewing this dissertation as a second reviewer.

Professor Dr. Martin Bremeier for helping me to settle in Abu Dhabi and reviewing this thesis as an external reviewer.

Dr. Shehadeh Masalmeh for guiding me in the right scientific way and his valuable comments and suggestions on our research papers.

A very special thank you to Sjaam Oedai for his invaluable advice and support. For sharing his immense knowledge of experimental research and for being a good friend in Abu Dhabi.

Gion Strobel for helping me whenever I struggled with the mysteries of $DuMu^x$.

Dr. Birger Hagemann for encouraging and inspiring me to write this dissertation.

Contents

List of Figures   ix

List of Tables   xiii

Glossary   xv

1 Introduction   1
- 1.1 Motivation 2
- 1.2 Production stages 3
- 1.3 Geology 3
- 1.4 Wettability 4
- 1.5 Low-salinity waterflooding 5
- 1.6 Outline 5

2 Scope & objectives   9
- 2.1 Experimental methodology 9
  - 2.1.1 Spontaneous imbibition 9
  - 2.1.2 Centrifuge method 11
  - 2.1.3 Unsteady state coreflooding 11
- 2.2 Numerical methodology 12
  - 2.2.1 Centrifuge simulation 12
  - 2.2.2 Coreflooding simulation 13

CONTENTS

    2.3   Material & equipment 14
    2.4   Experimental sequence & parameter 15

3 Literature review   **17**
    3.1   Buckley-Leverett equation 18
    3.2   Spontaneous imbibition experiments 20
    3.3   Coreflooding experiments 22
    3.4   Low-salinity mechanisms 26
    3.5   Summary & conclusions 32

I Experimental study   **35**

4 Preparation   **37**
    4.1   Fluid preparation 37
            4.1.1   Brine preparation 37
            4.1.2   Oil preparation 38
            4.1.3   Density & viscosity 39
            4.1.4   Interfacial tension 40
    4.2   Core preparation 42
            4.2.1   Core material 42
            4.2.2   Core cleaning 43
            4.2.3   Core selection 44
    4.3   Connate water saturation 45
            4.3.1   Core saturation 45
            4.3.2   Absolute brine permeability 45
            4.3.3   Nuclear Magnetic Resonance 48
            4.3.4   Primary brine drainage 50
            4.3.5   Effective oil permeability 52
            4.3.6   Core aging 52
            4.3.7   Unsteady state coreflooding samples 54

CONTENTS

	4.4	Summary & conclusions	54
5	**Spontaneous imbibition**	**57**
	5.1	Theory	57
	5.2	Spontaneous imbibition in carbonates	59
	5.3	Methodology	60
	5.4	Experimental conduction	62
	5.5	Spontaneous imbibition results	63
	5.5.1	Identical salinity of connate water and imbibing water	63
	5.5.2	Imbibition of a low saline brine into a system at higher salinity (A)	64
	5.5.3	Imbibition of a low saline brine into a system at higher salinity (B)	65
	5.5.4	Imbibition of a high saline brine into a system at lower salinity	65
	5.6	Contact angle	65
	5.7	Zeta potential	68
	5.8	Summary & conclusions	70
6	**Centrifuge method**	**75**
	6.1	Theory	75
	6.2	Methodology	77
	6.3	Analytical centrifuge solution	78
	6.4	Experimental conduction	84
	6.5	Centrifuge results	86
	6.5.1	Identical salinity of connate water and imbibing water	87
	6.5.2	Injection of a low saline brine into a system at higher salinity (A)	88

CONTENTS

    6.5.3   Injection of a low saline brine into a system at higher salinity (B) 88
    6.5.4   Injection of a high saline brine into a system at lower salinity 89
  6.6 Summary & conclusions 90

7 Coreflooding   93
  7.1 Theory 93
  7.2 Methodology 97
  7.3 Experimental conduction 98
  7.4 Coreflooding results 100
    7.4.1   Formation-water in secondary mode 101
    7.4.2   Sea-water in secondary mode 102
    7.4.3   Diluted-sea-water in secondary mode 102
  7.5 Summary & conclusions 103

II Numerical study   107

8 Numerical centrifuge simulation   109
  8.1 DuMu^x 109
    8.1.1   Spatial discretization 110
    8.1.2   Time discretization 113
    8.1.3   Newton's method 114
  8.2 DuMu^x centrifuge simulation 116
    8.2.1   Boundary conditions 116
    8.2.2   Fluid properties 119
    8.2.3   Hydraulic properties 119
    8.2.4   Flow equation 122
    8.2.5   Mass balance 124
    8.2.6   Time manager 124

CONTENTS

- 8.3 Cydar 125
- 8.4 Cydar centrifuge simulation 125
  - 8.4.1 Cydar input data 125
  - 8.4.2 Cydar boundary conditions 126
- 8.5 Simulation results 127
  - 8.5.1 History matching methodology 127
  - 8.5.2 Forced imbibition simulation 128
  - 8.5.3 Numerical capillary pressure 133
- 8.6 Summary & conclusions 134

9 Numerical coreflooding simulation   135

- 9.1 DuMu^x coreflooding simulation 136
  - 9.1.1 Boundary conditions 136
  - 9.1.2 Fluid properties 139
  - 9.1.3 Hydraulic properties 141
  - 9.1.4 Flow equation 143
  - 9.1.5 Molar balance 144
  - 9.1.6 Time manager 145
- 9.2 Cydar coreflooding simulation 145
  - 9.2.1 Cydar input data 145
  - 9.2.2 Cydar boundary conditions 146
- 9.3 Simulation results 146
  - 9.3.1 Absolute permeability matching 146
  - 9.3.2 History matching methodology 147
  - 9.3.3 Formation-water in secondary mode 150
  - 9.3.4 Sea-water in secondary mode 151
  - 9.3.5 Diluted-sea-water in secondary mode 157
  - 9.3.6 Relative permeability and capillary pressure . . 158
- 9.4 Summary & conclusions 160

CONTENTS

10 Summary & conclusions   **163**

Bibliography   **169**

A Experimental study appendix   **183**
    A.1 Spontaneous and forced imbibition results 183
    A.2 Unsteady state coreflooding results 183

B Numerical study appendix   **201**
    B.1 Source code excerpts - Centrifuge 201
    B.2 Source code excerpts - Coreflooding 206

C Ausführliche Zusammenfassung   **209**
    C.1 Einleitung 209
    C.2 Literaturrecherche 210
    C.3 Probenvorbereitung 212
    C.4 Spontaneous-imbibition Tests 213
    C.5 Zentrifugenversuche 214
    C.6 Kernflutungsexperimente 215
    C.7 Numerische Zentrifugensimulation 216
    C.8 Numerische Kernflutungssimulation 217
    C.9 Zusammenfassung 219

List of Figures

2.1　Experimental overview 10
2.2　Study methodology 16

3.1　Buckley-Leverett equation 19
3.2　Wettability alteration in limestones 28
3.3　Idealized Electrical double layer 30
3.4　Carbonate type impact on zeta potential 32

4.1　Brine density and viscosity 41
4.2　Crude oil properties 42
4.3　Core selection 44
4.4　Absolute and effective oil permeability measurement . 46
4.5　NMR core characterization 49
4.6　Primary drainage centrifuge method 51
4.7　Relative permeability comparison 53

5.1　Idealized spontaneous imbibition illustration 58
5.2　Co-current and counter-current imbibition scheme . . 60
5.3　Spontaneous and forced imbibition methodology . . . 61
5.4　Schematic Amott imbibition cell set-up 62
5.5　Spontaneous imbibition Group I 63
5.6　Spontaneous imbibition Group II 64

LIST OF FIGURES

5.7	Spontaneous imbibition Group III	66
5.8	Spontaneous imbibition Group IV	66
5.9	Spontaneous brine imbibition tests at 70°C	67
5.10	Contact angle measurements	67
5.11	Zeta potential measurement results	70
6.1	Primary drainage centrifuge sketch	76
6.2	Comparison of average and inlet water saturation	83
6.3	Imbibition centrifuge core holder set-up	85
6.4	Forced imbibition results Group I	88
6.5	Forced imbibition results Group II	89
6.6	Forced imbibition results Group III	89
6.7	Forced imbibition results Group IV	90
7.1	Capillary end effect	95
7.2	USS coreflooding methodology	97
7.3	Experimental coreflooding sketch	99
7.4	USS coreflooding In2b	101
7.5	USS coreflooding In4b	102
7.6	USS coreflooding In9b	103
8.1	Cell-centered finite volume method	111
8.2	Numerical centrifuge imbibition model	117
8.3	Capillary pressure model comparison	120
8.4	Comparison of initial and final history matching	129
8.5	Centrifuge history match In17	129
8.6	DuMu^x centrifuge simulation In17	130
8.7	Water saturation profile In17	131
8.8	Pressure profile In17	131
8.9	Capillary pressure profile In17	131
8.10	Analytical and numerical capillary pressure	133

LIST OF FIGURES

9.1	Numerical coreflooding model	137
9.2	Viscosity and density at 70°C	140
9.3	Numerical history match coreflooding In2b	151
9.4	Numerical history match coreflooding In4b	152
9.5	Relative permeability interpolation	152
9.6	DuMu$^\text{x}$ coreflooding simulation In4b	154
9.7	Water saturation profile In4b	155
9.8	Pressure profile In4b	155
9.9	Capillary pressure profile In4b	155
9.10	Numerical history match coreflooding In9b	157
9.11	Relative permeability curves	159
9.12	Capillary pressure curves	159
9.13	Field rate oil recovery	159
B.1	Dirichlet boundary implementation	203
B.2	Centrifugal acceleration implementation	204
B.3	Modified-P_c and Modified-kr law implementation	205
B.4	Neumann inlet boundary implementation	207
B.5	Neumann outlet boundary implementation	208

List of Tables

4.1	Brine composition	38
4.2	Arrhenius viscosity interpolation	40
4.3	Indiana limestone composition	43
4.4	Standardized permeability measurement example	48
4.5	Core properties	55
5.1	Zeta potential measurement results	71
5.2	Spontaneous imbibition results	73
6.1	Hyperbolic regression of centrifuge imbibition data	83
6.2	Centrifuge method results	91
7.1	USS coreflooding results	104
8.1	Newton's method	115
9.1	Single phase coreflooding model validation	147
A.1	Spontaneous and forced imbibition results In1	184
A.2	Spontaneous and forced imbibition results In2	185
A.3	Spontaneous and forced imbibition results In3	186
A.4	Spontaneous and forced imbibition results In4	187
A.5	Spontaneous and forced imbibition results In5	188

LIST OF TABLES

A.6　Spontaneous and forced imbibition results In7 189
A.7　Spontaneous and forced imbibition results In9 190
A.8　Spontaneous and forced imbibition results In10 191
A.9　Spontaneous and forced imbibition results In12 192
A.10 Spontaneous and forced imbibition results In13 193
A.11 Spontaneous and forced imbibition results In14 194
A.12 Spontaneous and forced imbibition results In15 195
A.13 Spontaneous and forced imbibition results In16 196
A.14 Spontaneous and forced imbibition results In17 197
A.15 USS coreflooding results In2b 198
A.16 USS coreflooding results In4b 199
A.17 USS coreflooding results In9b 200

Glossary

Abbreviations

BVI Irreducible water Bulk Volume

EDL Electrical double layer

EOR Enhanced oil recovery

FFV Free-Fluid Bulk volume

IFT Interfacial tension

NMR Nuclear Magnetic Resonance

PDI Potential determining ions

SCAL Special core analysis

USS Unsteady state

Roman letters

A	Cross-section	$[m^2]$
A_s	Pre-exponential factor	[-]
B	Hyperbolic least square constant	[-]
C	Hyperbolic least square constant	[-]
D	Hyperbolic least square constant	[-]

GLOSSARY

d_e	Pendant-drop maximal droplet diameter	$[m]$
d_s	Pendant-drop minimal droplet diameter	$[m]$
e	Modified-kr model parameter	$[-]$
E_α	Activation energy	$[J/mol]$
f	Modified-kr model parameter	$[-]$
G	Modified-kr model input parameter	$[-]$
g	Gravity	$[m/s^2]$
g_c	Centrifugal acceleration	$[m/s^2]$
$g_{c,rm}$	Average centrifugal acceleration	$[m/s^2]$
H	Modified-kr model input parameter	$[-]$
H_{PD}	Pendant-drop geometrical correction factor	$[-]$
I_{final}	Final pixel value (centrifuge method)	$[-]$
I_{init}	Initial pixel value (centrifuge method)	$[-]$
I_{insitu}	In-situ pixel value (centrifuge method)	$[-]$
k	Absolute permeability	$[m^2]$
$ko_{Sw,c}$	Effective oil permeability at connate water saturation	$[-]$
kr_o	Relative oil permeability	$[-]$
kr_o^{max}	Endpoint oil relative permeability	$[-]$
kr_w	Relative water permeability	$[-]$
kr_w^{max}	Endpoint water relative permeability	$[-]$
$kw_{So,rm}$	Effective water permeability at remaining oil saturation	$[-]$
M	Molar mass	$[kg/kmol]$

GLOSSARY

m	Mass	$[kg]$
m_{dry}	Core dry weight	$[kg]$
m_{wet}	Core wet weight	$[kg]$
n	Amount of substance	$[kmol]$
N_B	Bond number	$[-]$
N_C	Capillary number	$[-]$
P_c	Capillary pressure	$[Pa]$
P_{atm}	Atmospheric pressure	$[Pa]$
P_{ci}	Capillary inlet pressure	$[Pa]$
P_c	Capillary pressure	$[Pa]$
P_{inlet}	Inlet pressure	$[Pa]$
P_{outlet}	Outlet pressure	$[Pa]$
P_o	Oil phase pressure	$[Pa]$
P_w	Water phase pressure	$[Pa]$
PF	Pixel factor (centrifuge method)	$[-]$
q_w	Oil phase source/sink term	$[kg/(s \cdot m^3)]$
q_w	Water phase source/sink term	$[kg/(s \cdot m^3)]$
q_{lab}	Volumetric laboratory injection rate	$[cm^3/min]$
R	Universal gas constant	$[J/molK]$
r_m	Distance of the center of rotation to the center of core	$[m]$
r_p	Pore radius	$[\mu m]$
r_{inlet}	Distance of the axis of rotation to the core inlet	$[m]$

GLOSSARY

r_{outlet}	Distance of the axis of rotation to the core outlet	[m]
RPM	Revolution per minute	[-]
So	Oil saturation	[-]
So_r	Residual oil saturation	[-]
Sw	Water saturation	[-]
Sw_c	Connate water saturation	[-]
Sw_{av}	Average water saturation	[-]
Sw_{DS}	Dean-stark validated water saturation	[-]
Sw_e	Normalized water saturation	[-]
Sw_{inlet}	Inlet water saturation	[-]
Sw_{outlet}	Inlet water saturation	[-]
T	Temperature	[K]
t	Time	[s]
T_2	Spin-spin relaxation time	[μs]
v	Flow velocity	[m/s]
V_C	Core sample volume	[m^3]
v_f	Field flow velocity	[feet/day]
v_o	Advective volumetric oil velocity	[m/s]
V_P	Pore volume	[m^3]
V_w	Water volume	[cm^3]
v_w	Advective volumetric water velocity	[m/s]
V_o^{cap}	Captured oil (centrifuge method)	[cm^3]

GLOSSARY

V_o^{init}	Initially added oil (centrifuge method)	$[cm^3]$
L	Core length	$[m]$
r	Radius	$[m]$

Greek symbols

α	Forbes geometrical correction factor	[-]
β	Forbes geometrical correction factor	[-]
δ	Surface relaxivity	$[\mu m/s]$
$\Delta \rho$	Density difference	$[kg/m^3]$
ϵ	Modified-kr model input parameter	[-]
μ	Dynamic viscosity	$[Pas]$
μ_i	Oil viscosity	$[Pas]$
μ_w	Water viscosity	$[Pas]$
ω	Angular velocity	$[rad/s]$
ϕ	Porosity	[-]
ρ_o	Oil density	$[kg/m^3]$
ρ_w	Water density	$[kg/m^3]$
σ	Interfacial tension	$[N/m]$
σ_{ow}	Oil water interfacial tension	$[J/m^2]$
θ	Contact angle	[-]
ΔP	Differential pressure	$[bar]$

1

Introduction

In 2013, a cooperation between the Petroleum Institute Abu Dhabi and the Institute of Petroleum Engineering Clausthal was established. The cooperation targets the exchange of ideas, concepts and the creation and transmission of knowledge in the framework of an international research partnership. As part of this cooperation, a comprehensive experimental and numerical study on low-salinity effects in limestone samples was developed. The vast majority of the research project was completed during a three and a half years long research stay at the Petroleum Institute Abu Dhabi.

The study fundamentally benefited from the unique research opportunities at the Petroleum Institute Abu Dhabi. Particularly, the newly established ADNOC Research and Innovation Center provides an exceptional Enhanced oil recovery specialized collection of state-of-the-art research equipment. Furthermore, the study profited from the short link to the Enhanced oil recovery division of the Abu Dhabi National Oil Company. The experimental findings were presented to an international audience at the 2018 Society of Core Analysis Symposium in Trondheim, Norway [34].

The capabilities of the Institute of Petroleum Engineering Clausthal contributed to the numerical analysis of the experimental data. Based on several years of experience on the open-source C++ simulator DuMu^x, the Institute assisted in the development of an independent numerical centrifuge and coreflooding model. A description of the implemented mathematical and numerical model approach is (going to be) published in the Journal of Petroleum Science and Engineering [35].

1.1 Motivation

The 2019 United Nations World Population Prospects forecasts a world population of more than 9.7 billion people in the year 2050 [102]. Related to this significant population growth, the World Energy Council [40] predicts a total primary energy supply increase of 27 to 61 %. Although the development and expenditures on renewable energy systems significantly change the energy production mix, the dependency on the fossil fuels coal, oil and gas remains high. It is expected that approximately one-quarter of the primary energy consumption in 2050 will still be supplied by the combustion of hydrocarbons. Besides the development of alternative concepts, especially the transport sector will predominantly depend on combustion engines, which emphasizes the continuous importance of the Oil & Gas energy sector.

Since it is generally believed that the majority of the existing oil reservoirs are discovered, the oil industry increasingly focuses on enhancing field recovery factors to meet the unabated oil demand. By currently holding 7 % of the worldwide oil reservoirs, Abu Dhabi plays a major role in maintaining the required oil supply [84]. While the current worldwide average recovery factor amounts approximately 35 %, the Abu Dhabi National Oil Company targets ambitious and challenging recovery factors of up to 70 % [100].

The huge hydrocarbon reservoirs of the Middle East are typically located inside carbonate formations. Albeit carbonate reservoirs are believed to contain 60 % of the world's oil reserves, several factors such as complex rock micro-structures, heterogeneous porosity and permeability distributions and unfavorable wettability conditions cause an inefficient oil production [95]. In order to achieve significantly higher recovery rates, it is therefore inevitable to invest in improved oil production concepts and technologies.

1.2 Production stages

The production of an oil field is typically divided into three recovery stages. During the primary recovery, the oil production benefits from the naturally existing reservoir drive. Typical examples are an aquifer, solution-gas, gas cap and/or rock and fluid expansion drive. Depending on the primary recovery drive and reservoir properties, 5 to 15 % oil can be recovered without the application of external energy [92].

Once the primary recovery decreases, oil production is typically maintained by the introduction of external energy. The secondary recovery stage focuses on the reservoir pressure maintenance and volumetric sweep efficiency increase due to the injection of immiscible fluids. Most secondary oil production stages are represented by conventional waterfloodings.

Tertiary recovery refers to the last production stage of an oil field, in which the oil recovery is stimulated by the injection of a fluid, which usually does not occur inside the reservoir. Examples are the injection of miscible and immiscible gases, chemical floodings, thermal application and other methods. Tertiary recovery technologies often conflict with economic interests, as tertiary oil production methods cause incremental production costs.

1.3 Geology

The vast majority of Abu Dhabi's oil reservoirs are located inside the Thamama Group (Lower Cretaceous) and the Arab Formation (Upper Jurassic). The porosity typically ranges between 20 to 30 %, while the permeability locally exceeds $100\,md$ [48, 81].

As a result of carbonate reservoir genesis, the spatial porosity and permeability distribution within carbonates formation are strongly fluctuating. During the deposition, grain size, packing and sorting determine the fundamental rock structure. Typical carbonate depositional environment range from tidal flats to deep-water basins. The subsequent diagenesis includes several geological processes that affect the rock set-up. Compaction and cementation reduce the original porosity, while in contrast, chemical dissolution increases the void space. Thereby the dissolution of calcite can lead to the creation of new pores, which are typically not connected to original void space [3]. Besides primary and secondary porosity, fractured porosity is known as a third possible porosity type in carbonates. Caused by high mechanical stress, the arising joints (fractures) lead into a locally unrestricted flow. Typical consequences are matrix bypassing, early water breakthrough and/or gas coning [95].

1.4 Wettability

Besides the strong permeability and porosity heterogeneity of carbonates, the typically intermediate to oil-wet rock surface complicates an efficient oil production. In general, the concept of wettability describes the tendency of a fluid to adsorb on or detach from a surface under the presence of at least a second immiscible fluid [93]. After the hydrocarbon accumulation inside the reservoir formation, a chemical and physical equilibrium between the initial reservoir fluid, hydrocarbons, impurities and rock arises over a time period of millions of years [93]. Depending on the fluid properties and mineralogy of the reservoir, either water, oil or a mixture of both fluids tends to wet the rock surface. While the majority of sandstone reservoirs are assumed to be water-wet, carbonate reservoirs are characterized by an intermediate/oil-wet wettability. As a result, injected water typically bypasses the surface adsorbed oil, which hence causes an inefficient oil recovery.

As one of the primary purposes of tertiary oil recovery, Enhanced oil recovery (EOR) applications are conducted in the hope of altering the reservoir wettability. By shifting the initially intermediate or oil-wet wettability to more water-wet conditions, it is assumed that initially adsorbed oil can be re-mobilized [95].

1.5 Low-salinity waterflooding

Besides the complicated geology and unfavorable wettability conditions, the Middle East carbonate reservoirs are furthermore characterized by a high saline and high temperature environment. In regard to the typically light crude oil containing formations, the application of several Enhanced oil recovery methods is reasonable. As an alternative to the more expensive chemical EOR methods such as polymer, surfactant and/or alkaline floodings, this thesis evaluates the potential of low-salinity waterflooding to increase oil recovery.

The work of Jadhunandan [55, 94] (1991) was one of the first publications, in which a correlation between brine composition and oil recovery was described. In general, the concept of low-salinity water injection targets the reservoir wettability alteration towards stronger intermediate or water-wet conditions by the injection of a desalinated and/or modified injection brines.

Although low-salinity is generally linked to wettability alteration, the involved physical and chemical mechanisms are still controversial. Starting from five published low-salinity publications in 2007, the amount of proposed low-salinity mechanisms increased to already seventeen in 2014 [94]. Besides brine and oil properties, particularly the reservoir type significantly impacts the physics of low-salinity effects. Consequently, the proposed low-salinity mechanisms fundamentally differ for sandstone, chalk, limestone and dolomite samples.

The current research findings on low-salinity effects in limestones are summarized in Chapter 3. The literature review points out that the impact of low-salinity waterflooding on oil recovery is still not well understood. This thesis therefore develops, conducts and analyzes a comprehensive low-salinity study on limestone samples to evaluate the impact of low-salinity waterflooding on oil recovery.

1.6 Outline

The thesis is divided into ten chapters. In line with the current status of low-salinity research on limestone, Chapter 2 develops and

1.6 OUTLINE

formulates the scope, objectives and methodology of the conducted study.

Chapter 3 initially introduces the concept of low-salinity waterflooding at the example of the Buckley-Leverett solution. A comprehensive literature review summarizes the reported spontaneous imbibition and corefloodings experiments on limestone samples. Furthermore, the proposed low-salinity mechanisms on carbonates are introduced.

The subsequent chapters are divided into an experimental and numerical part. In the first place, Chapter 4 describes the experimental preparation of the study. The chapter emphasizes that a careful experimental preparation is crucial to obtain high-quality research data. Besides the core and fluid treatment, the section pictures the establishment of the initial sample conditions.

Chapter 5 describes the theory, experimental conduction, results and conclusions of the spontaneous imbibition section. The spontaneous imbibition experiments include fourteen test samples to evaluate the impact of connate and imbibing water composition impact on spontaneous oil recovery. Additionally, zeta potential measurements are conducted in order to evaluate the involved low-salinity mechanisms.

The principle of the centrifuge method is initially explained at the example of primary drainage (Chapter 6). The imbibition experiments are analytically corrected by the combination of Forbes First solution and a hyperbolic fit of the acquired forced imbibition data.

The three conducted unsteady state corefloodings are described in Chapter 7. Besides a description of the experimental conduction and results, Chapter 7 includes a general introduction of the unsteady state coreflooding technique.

The numerical part of this thesis includes the development of a centrifuge and coreflooding simulation within the C++ open-source software DuMu$^\text{x}$. Chapter 8 derives a two-phase numerical centrifuge model. The development of the numerical model includes the implementation and evaluation of suitable boundary conditions, centrifu-

1.6 OUTLINE

gal force and hydraulic property adaption. Furthermore, the chapter validates the developed centrifuge model against the commercial Cydar software.

Chapter 9 completes the numerical work by the development of a two-phase-three-component coreflooding simulation. Due to the introduction of a salt tracer component, the proposed coreflooding model allows the simulation of viscosity changes and low-salinity effects. In accordance with the centrifuge model, the presented coreflooding model is validated against Cydar.

Finally, the main findings of the study are summarized in Chapter 10. The spontaneous imbibition, centrifuge method and displacement results are attached in Appendix A. An impression of the numerical work is provided in Appendix B, where excerpts of the source code illustrate the C++ implementation of the proposed mathematical centrifuge and coreflooding model.

2
Scope & objectives

The implemented low-salinity study combined spontaneous imbibition, centrifuge method and USS corefloodings to investigate the impact of brine composition on oil recovery. Each method is suitable to cover a specific aspect of the complex reservoir oil recovery processes [72]. Furthermore, zeta potential measurements were conducted to gain a better understanding of low-salinity mechanisms in limestones. The study is completed by the development of a numerical centrifuge and a coreflooding model to validate and history match the experimental data.

2.1 Experimental methodology

2.1.1 Spontaneous imbibition

Spontaneous imbibition experiments are a simple and widely used method to screen the potential of imbibing fluids to recover oil spontaneously. In regard to carbonate reservoirs, in which highly permeable and conductive fractures surround low permeable matrix systems, spontaneous imbibition is believed to be an important recovery mechanism [74]. Although spontaneous imbibition tests are a frequently reported experimental procedure, it remains questionable

2.1 EXPERIMENTAL METHODOLOGY

if the experiments are suitable to evaluate the potential of improved oil recovery applications [69].

The conducted spontaneous imbibition test sequence included eight different connate water and imbibing brine combinations. In line with the literature, the study initially evaluated the capability of a medium saline (Sea-water) and a low saline (Diluted-sea-water) brine to spontaneously recover oil from a system at higher salinity. To improve the understanding of spontaneous imbibition physics, the study additionally examined the connate water composition impact on oil recovery. Therefore, a sequence of spontaneous imbibition tests was conducted, in which the connate water and imbibing water had an identical composition. Moreover, low saline connate water was combined with a high saline imbibing water. All tested connate water and imbibing combinations are summarized in Figure 2.1.

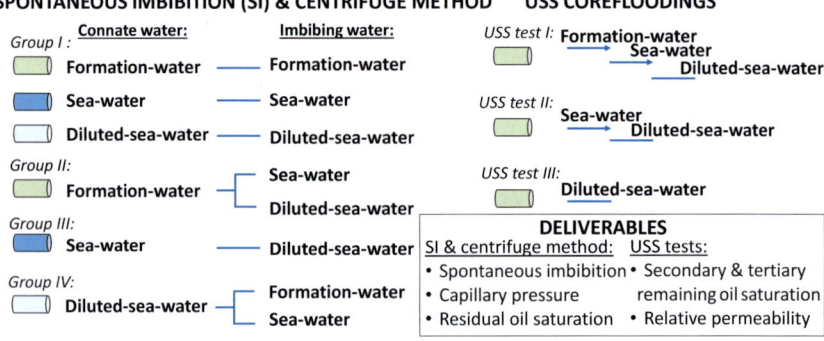

Figure 2.1: Experimental overview - The overview summarizes the spontaneous imbibition, centrifuge method and unsteady state corefloodings experiments.

As part of the spontaneous imbibition chapter, zeta potential measurements were conducted to investigate the involved mechanisms of the spontaneous oil recovery. A comprehensive overview of the proposed low-salinity mechanisms is provided in Chapter 3. The scope and objectives of the spontaneous imbibition tests can be summarized as follows

2.1 EXPERIMENTAL METHODOLOGY

- Evaluation of the Sea-water and Diluted-sea-water potential to spontaneously recover oil in comparison to the high saline Formation-water,

- Connate water and imbibing brine compositions prerequisites to obtain spontaneous oil recovery,

- Zeta potential and contact angle measurements to evaluate possible low-salinity effects.

2.1.2 Centrifuge method

In general, the centrifuge method is applied to derive capillary pressure drainage and imbibition data. Since the arising water-oil replacement process of the centrifuge method is gravity stable and hence does not interfere with viscous fingering, the method estimates the (true) residual oil saturation [72].

After completing the spontaneous imbibition tests, the samples were mounted inside a centrifuge and drained to the residual oil saturation So_r. Besides determining the true residual oil saturation, the derived capillary pressure data reflects the brine composition impact on wettability. The combination of spontaneous imbibition and centrifuge method furthermore evaluated the cohesion of spontaneous imbibition and residual oil saturation. The primary objectives of the experimental centrifuge method part are

- Impact of connate water and imbibing brine composition on wettability and residual oil saturation,

- Cohesion of spontaneous and forced imbibition.

2.1.3 Unsteady state coreflooding

After completing the spontaneous and forced imbibition tests, the experimental part of the study was continued by the conduction of three unsteady state (USS) core floodings. Thereby, each brine was tested in secondary and tertiary injection mode. While the sec-

2.2 NUMERICAL METHODOLOGY

ondary injection mode refers to the water injection at connate water saturation, the tertiary injection reflects the water injection at a mature recovery stage. In comparison to spontaneous imbibition and centrifuge method, USS corefloodings are believed to represent an oil-water displacement process under more field-realistic conditions. The objectives of the coreflooding experiments are as follows

- Impact of brine composition on secondary and tertiary oil recovery,
- Comparison of spontaneous imbibition, centrifuge method and corefloodings method results.

2.2 Numerical methodology

2.2.1 Centrifuge simulation

The main objectives of the centrifuge method and unsteady state corefloodings experiments are listed in the previous sections. In order to validate and history match the acquired experimental data, Special core analysis (SCAL) studies typically include analytical and numerical methods.

However, the unavoidable interference of capillary pressure and relative permeability data significantly complicates the data analysis of SCAL experiments. Since most analytical approaches neglect capillary pressure, analytically obtained parameters such as the residual oil saturation, relative permeability curves, e.g., are often distorted [62, 63].

As an alternative to an analytical approach, numerical simulations allow the consideration of capillary pressure. Cydar, Porlab, Scores and Sendra are examples for SCAL specialized numerical simulation programs [60]. However, as a typical drawback of commercial software, the numerical and mathematical model implementation remains unclear to the end-user.

This work therefore aims to develop an independent and transparent numerical centrifuge and coreflooding simulation. The nu-

2.2 NUMERICAL METHODOLOGY

merical models were thereby built on top of the open-source DuMu$^\text{x}$ simulator to avoid a tedious start from scratch. The work provides a detailed mathematical description of the implemented flow model, fluid system, hydraulic properties and boundary formulation. Furthermore, the presented numerical centrifuge and coreflooding models were validated against the commercial Cydar software.

Initially, the numerical centrifuge simulations evaluate the physical plausibility of the experimentally acquired data by conducting numerical history matches. Since the arising centrifuge history match is predominantly impacted by capillary pressure, the numerical centrifuge simulation is not used to derive relative permeability data. The main objectives of the numerical centrifuge part are

- Development and implementation of a numerical centrifuge model in DuMu$^\text{x}$,

- History matches of the experimental centrifuge data. Derivation of numerical capillary pressure.

2.2.2 Coreflooding simulation

After determining the spontaneous imbibition behavior, capillary pressure and residual oil saturation of the eight connate water and imbibing water combinations, the USS coreflooding data was used to derive relative permeability numerically. In line with the centrifuge experiments, the numerical coreflooding simulations were furthermore conducted to evaluate the physical plausibility of the experiments. The main objectives of the numerical coreflooding chapter are

- Development and implementation of a numerical coreflooding model in DuMu$^\text{x}$,

- History matching of the experimental coreflooding data. Derivation of relative permeability data.

2.3 Material & equipment

As summarized in the literature review, the low-salinity research on limestones currently focuses on two approaches to increase oil recovery: The introduction of multivalent ions and/or the reduction of the total ionic strength. In line with the literature, the implemented study tested three different brine compositions to investigate the impact of brine composition on oil recovery. Besides Formation-water as a high saline brine (salinity of $183.4\,g/l$), Sea-water represented a medium saline (salinity of $43.8\,g/l$) and sulfate-rich (sulfate concentration of $3.5\,g/l$) imbibing brine. Furthermore, (100 times) Diluted-sea-water (salinity $0.44\,g/l$) was selected to represent an imbibition process, in which the system's salinity is significantly reduced.

The oil recovery study used a low viscous Middle East dead crude oil. A detailed description of the oil and brine property measurements is provided in Chapter 4.

The majority of the Middle East's oil reservoir is located in carbonate limestone formations. These formations are characterized by complex rock micro-structures, heterogeneous porosity and permeability distributions as well as unfavorable wettability conditions [17, 81, 95]. In regard to the limited availability and the high heterogeneity of reservoir cores, this study used Indiana limestone outcrop samples. As a fundamental benefit of outcrop cores, the samples provide homogeneous and comparable porosity and permeability properties.

The experimental part of this work was conducted in the ADNOC Research Innovation Center. The research facility provides a comprehensive and state-of-the-art collection of petroleum engineering specialized research equipment. Besides fluid and core material property measurement systems, this study especially benefited from in-situ saturation monitoring centrifuges and an automatized multi-phase coreflooding system.

2.4 Experimental sequence & parameter

After formulating the objectives and selecting the materials, the experimental sequence and parameter definition specifies the study schedule. The six milestones of the study are summarized in Figure 2.2, in which the numbering corresponds to the chapter numbering of this thesis.

Bullet point 4 describes the experimental preparation of the study. Initially, the brine is synthesized based on an ionic composition analysis. The brines and oil were then treated to remove undesired contaminations. Viscosity and density properties were measured within a temperature range of 20 to 60°C and interpolated to the test conditions of 70 °C. The core preparation included core cutting, core cleaning, helium porosity and nitrogen permeability measurements. Finally, the initial water saturation was established by saturating, draining and aging the limestone samples. The establishment of the initial core conditions was monitored by measuring the effective oil permeability before and after aging.

Without the application of back-pressure, brine and crude oil vaporization limits the maximum spontaneous imbibition test conditions to a maximum temperature of 70°C. To ensure the comparability of the experiments, the spontaneous imbibition, centrifuge method and USS displacement experiments were conducted at a consistent temperature of 70°C (Figure 2.2, bullet point 5, 6, 7). While the fourteen spontaneous imbibition tests were conducted over a time period of 30 $days$, the nine consecutive centrifuge spin steps of 1000 - 7500 RPM corresponded to a total centrifuge run duration of 11 - 12 $days$. The USS corefloodings were realized at a confining pressure of 107 bar and back-pressure of 14 bar. Depending on the injection pattern, the duration of the corefloodings varied between 2 to 5 $days$.

2.4 EXPERIMENTAL SEQUENCE & PARAMETER

4.1) Fluid preparation
4.1.1: Brine preparation:
- Brine synthesis & filtration

4.1.2: Oil preparation:
- Centrifuge treatment to clean oil
  − 5000 RPM, 30 min

4.1.3: Viscosity & density :
- Measurement range:
  − 20 - 60°C, atm
- Interpolation to 70°C

4.1.4: IFT measurements:
- Pendant Drop
  − Room temperature, atm

4.2) Core preparation:
4.2.1: Core cutting:
- 5 cm x 3.7 cm

4.2.2: Core cleaning:
- 10 PVs flush-through cleaning
- Methanol injection, room temperature

4.2.3: Core selection:
- Helium porosity & nitrogen permeability
- CT-Scan imaging
- Core selection

4.3) Establishment of Sw_c
4.3.1 Core saturation:
- Vacuum evacuation; connate water saturation
  − 140 bar, 48 hours

4.3.2 Brine permeability
- Absolute brine permeability
  − Room temperature, 7 bar backpressure (BP)

4.3.3 T2 NMR test:
- NMR tests at 100% brine saturation
- Pore size distribution, NMR based Sw_c reference value

4.3.4 Primary brine drainage:
- Centrifuge method
  − Single-speed, 12000 RPM
  − 70°C, 48 hours

4.3.5 Core aging:
- Measuring of kr_o at Sw_c before aging:
  − 45°C, 7 bar backpressure
- Core ageing:
  − up to 30 days, 90°C, 7 bar backpressure
- Measuring of kr_o at Sw_c after aging:
  − 45°C, 7 bar BP

5) Spontaneous imbibition
- Eight water and imbibing water combinations:
  − 70°C, atm, 30 days
- Zeta potential measurements:
  − Limestone brine suspension, 25 - 55°C, atm

6) Centrifuge method
- Forced imbibition using the centrifuge method:
  − At 70°C, 9 multi-speed steps, each step 30 hours: 1000, 1280, 1650, 2110, 2700, 3470, 4450, 5700, 7500 RPM

7) Coreflooding
- USS corefloodings:
  − Rate bumping; 0.05, 0.02, 1, 3, 5 cm^3/min
  − Secondary and tertiary brine injection
  − 70°C
  − Confining pressure 107 bar, 14 bar back-pressure
- Dean-Stark-extraction:
  − Vaporized toluene as solvent

8) Numerical centrifuge simulation
- Development of a DuMu^x centrifuge model:
  − To history match experimental data
  − Validation against Cydar software
  − Derivation of numerical capillary pressure
  − 100 x 1 grids

9) Numerical coreflooding simulation
- Development of a DuMu^x coreflooding model:
  − To history match experimental data
  − Validation against Cydar software
  − Derivation of numerical relative permeability
  − 100 x 1 grids

Figure 2.2: Study methodology - The study overview specifies the schedule, parameters and experimental sequence of the implemented low-salinity study.

3
Literature review

The concept of low-salinity injection into sandstones has been extensively investigated since the early 90s. Although the mechanisms of low-salinity effects are still not well understood, it is widely accepted that low-salinity injection has a positive impact on oil production [94]. During the early state of low-salinity research on carbonates, the literature initially focused on chalk formations. The research group of Austad et al. [98, 111] conducted comprehensive low-salinity studies on Stevns Klint chalk outcrops, demonstrating that besides the total salinity, especially multivalent ions such as magnesium, calcium and sulfate affect the sample's wettability and oil recovery. In regard to the impact on low-salinity effects, the literature often summarizes multivalent ions (magnesium, calcium, sulfate, e.g.) as potential determining ions (PDI).

As a result of the promising experimental results on chalk samples, low-salinity effects in carbonates are often linked to concentration changes of potential determining ions. However, considering the different depositional environment and diagenetic history of chalk, limestone and dolomite reservoirs, the findings of the low-salinity research should be classified according to the corresponding carbonate type. While chalk reservoirs are the result of the compaction and cementation of marine organisms, limestone reservoirs are a mixture of carbonate mud and fragments of biogenic materials [57]. Although

3.1 BUCKLEY-LEVERETT EQUATION

both reservoir types are composed of calcium carbonate, chalk has a finer texture and is furthermore purely biogenic. Since the chalk pore surface is typically much larger compared to the limestone surface, the reactivity of chalk towards ions is supposed to be higher. Therefore, it is questionable if the suggested chalk low-salinity mechanisms are applicable to limestones samples [98].

3.1 Buckley-Leverett equation

Before reviewing the literature on low-salinity effects on limestones, this section initially introduces the theoretical (analytical) concept of secondary and tertiary low-salinity waterflooding.

The Buckley-Leverett equation is a conservation equation which describes the immiscible two-phase displacement at one-dimension. Although the Buckley-Leverett solution is based on several assumptions, it is a powerful method for the analytical illustration of a waterflooding displacement process. Besides assuming incompressibility and immiscibility, the Buckley-Leverett theory neglects capillary and gravity forces. Furthermore, the fluid flow obeys Darcy's law. A comprehensive mathematical derivation of the Buckley-Leverett solution can be found in the work of Dake [28].

The Buckley-Leverett solution of a conventional waterflooding is drawn in Figure 3.1, a (dashed line). In this idealized displacement case, the porosity, permeability and initial water saturation are homogeneously distributed. Once water is injected from the left-hand side, the arising water front pushes the oil towards the producer. Ahead of the displacement front, the water remains immobile due to the non-existing relative water permeability. Consequently, before the water breakthrough, only oil is produced. The oil production rate thereby equals the water injection rate. As the water-front reaches the producer, oil and water are initially co-produced. With increasing production time, the oil saturation approaches residual oil saturation.

Independent from the physical and chemical mechanisms, the most generalized low-salinity simulation approach includes the in-

3.1 BUCKLEY-LEVERETT EQUATION

terpolation between an oil-wet and water-wet relative permeability set. In this case, an analytical low-salinity solution can be derived based on Pope's [86] Buckley-Leverett polymer flooding approach. The analytical Buckley-Leverett derivation is possible for the secondary and tertiary low-salinity injection mode.

In the first place, Figure 3.1, a sketches the extended Buckley-Leverett solution of a secondary injection mode. As a result of the definition of an oil and water-wet permeability set, the arising Buckley-Leverett solution is characterized by two water-fronts: The first shock marks the transition between connate water and connate bank, while the subsequent shock separates the connate bank and the injected low-salinity water.

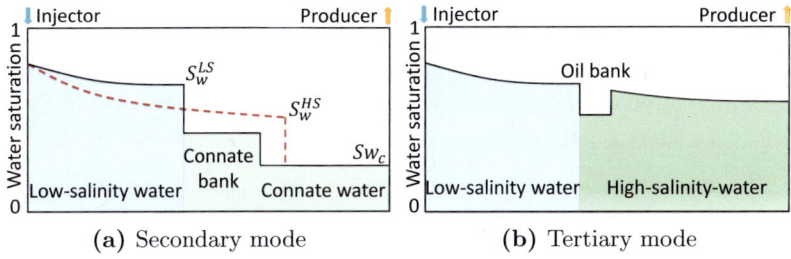

(a) Secondary mode (b) Tertiary mode

Figure 3.1: Buckley-Leverett equation - Comparison of the Buckley-Leverett solution of an ordinary waterflooding (Figure 3.1, a, dashed line), secondary mode low-salinity injection (a) and tertiary mode low-salinity injection (b) (redrawn after [29]).

In accordance with conventional waterflooding, the fluid extraction at the producer is initially limited to the oil phase. Once the first shock reaches the producer, the oil production drops to a constant value somewhere between the connate water saturation Sw_c and the high-salinity shock S_w^{HS}. The connate bank between the first and second shock is caused by the displacement of the high-saline water, which correspondingly results in the additional displacement of oil. Consequently, the breakthrough of the subsequent second shock marks the arrival of low-salinity water. Compared to the ordinary waterflooding (Figure 3.1, a, dashed line), the low-salinity displacement process is characterized by a slower front migration and hence

3.2 SPONTANEOUS IMBIBITION EXPERIMENTS

by a delayed water breakthrough. The high water saturation of the second front (Figure 3.1, S_w^{LS}) indicates a reduction of the residual oil saturation.

The Buckley-Leverett solution of tertiary low-salinity injection is plotted in Figure 3.1, b. In the initial stage, the producer extracts at a high water-cut. Due to the injection of a low-salinity water, the relative permeability is altered towards stronger water-wet conditions. As a result of the altered relative permeability (cf. Figure 4.7), additional oil is mobilized, which results in the formation and production of an oil bank [29, 56, 86].

3.2 Spontaneous imbibition experiments

A common and widely applied technique to screen the impact of imbibing fluids on oil recovery is spontaneous imbibition experiments. The subsequent section summarizes the literature on spontaneous imbibition tests on limestone samples.

The low-salinity research focuses on two approaches to increase oil production in limestones. First, it is believed, that the modification of potential determining ion concentration can lead to the replacement of the acid oil components from the positively charged limestone surface. Secondly, it has been shown that the injection of very low saline brines can improve oil recovery [89].

The potential of sulfate ions to increase spontaneous oil production in two Middle East limestone reservoir cores was demonstrated by Strand et al. [52, 98]. The spontaneous oil recovery of synthetic Sea-water was approximately 15 % higher compared to the imbibition of sulfate-free reference Sea-water.

A similar trend was shown in the work of Lighthelm et al. [61], in which three spontaneous imbibition experiments on Middle East limestone reservoir cores were conducted. Compared to the reference imbibing brine (sulfate concentration of $1.8\,g/l$), the two sulfate-enriched test brines (sulfate concentration of $4.1\,g/l$ and $9.5\,g/l$) resulted into an approximately 5 % and 10 % higher spontaneous oil recovery.

3.2 SPONTANEOUS IMBIBITION EXPERIMENTS

While the studies, as mentioned above, focused on the impact of sulfate concentration on spontaneous imbibition behavior, other research groups additionally investigated the impact of the total ionic strength on oil recovery. On the one hand, the work of Nasaralla et al. [82] confirmed the potential of sulfate-rich Sea-water to spontaneously displace oil from limestone cores, on the other hand, it was simultaneously demonstrated that the imbibition of 10 times diluted Sea-water has higher potential to displace oil spontaneously. In secondary mode, the conducted spontaneous imbibition on limestone reservoir cores resulted into an approximate average spontaneous oil recovery of 5 % for Formation-water as imbibing water, 9 % spontaneous oil recovery for Sea-water as imbibing water and 17 % spontaneous oil recovery for 10 times diluted Sea-water as imbibing water. In the second sequence of spontaneous imbibition tests, Nasaralla et al. [82] furthermore evaluated the potential of Sea-water and 10 times diluted Sea-water to increase spontaneous oil recovery in tertiary imbibition mode. After the imbibition of Formation-water, the tertiary imbibition of Sea-water caused an additional oil recovery of 5 %. Exchanging the tertiary imbibing fluid from Sea-water to 10 times diluted Sea-water caused an additional oil recovery of approximately 2 %.

The observations of Nasaralla et al. [82] are in accordance with the study of Romanuka et al. [89], in which Middle East limestone reservoir cores were used for spontaneous imbibition tests. In a first spontaneous imbibition test sequence, the tertiary imbibition behavior of a highly diluted brine and a potential determining ion enriched brine was compared. After the secondary imbibition of the highly diluted brine, the tertiary imbibition of the PDI enriched test brines did not increase the spontaneous oil recovery. Romanuka et al. [89] therefore concluded that spontaneous imbibition in limestone is promoted by the reduction of the total salinity rather than the modification of the PDI concentration. The work was completed by a second and third sequence of spontaneous imbibition tests, in which the highly diluted brine (salinity of $1\,g/l$ of sodium chloride and insignificant sulfate concentration) caused the overall highest spontaneous oil recovery in the range of 15 to 35 %.

3.3 COREFLOODING EXPERIMENTS

Zhang and Sarma [107] confirmed the potential of Diluted-seawater to increase spontaneous oil recovery in tertiary mode. The primary spontaneous imbibition of Sea-water into a Middle East limestone reservoir initially caused a spontaneous oil recovery of roughly 4.6 %. As a result of the imbibition brine exchange from Sea-water to 40 times diluted Sea-water, an additional oil recovery of 18.4 % was recorded.

In accordance with Zhang and Sarma [107], Al Harrasi et al. [5] conducted several spontaneous imbibition experiments on Middle East carbonate reservoir samples. Using Formation-water as imbibing brine, 9 % oil was spontaneously recovered. The exchange of the imbibing brine to 2 times diluted Formation-water increased the spontaneous oil recovery to 17.2 %. The subsequent imbibition of 100 times diluted Formation-water caused a final oil spontaneous recovery of 25 %. The spontaneous imbibition tests of Al Harrasi et al. [5] were showing a consistent trend. Exchanging the initial imbibing brine by a higher diluted imbibing brine caused additional spontaneous oil recovery.

This section summarized the results of six spontaneous imbibition studies on limestone samples. In secondary imbibition mode, highly diluted imbibing brines indicate a significant potential to cause spontaneous oil recovery. Furthermore, in the case of tertiary spontaneous imbibition mode tests, additional spontaneous oil recovery was recorded when the primary (medium saline) imbibing water was exchanged by a higher diluted imbibing water. In line with the conclusion of Romanuka [89], the literature review demonstrates that reducing the total ionic strength of the imbibing water is a promising approach to promote spontaneous imbibition in limestone samples.

3.3 Coreflooding experiments

Although spontaneous imbibition tests are widely used to screen the applicability of injection fluids, they do not necessarily reveal evidence about improved oil recovery [69]. To study the impact on remaining and residual oil saturation, displacement experiments such

3.3 COREFLOODING EXPERIMENTS

as corefloodings or centrifuge experiments are usually conducted. While there is a gap of low-salinity centrifuge data in the literature, comprehensive low-salinity studies using the unsteady state coreflooding method are available. This section summarizes the published literature on unsteady state corefloodings on limestone samples.

Al-Attar et al. [4] compared the oil recovery efficiency of several brine compositions in secondary injection mode. Besides evaluating the impact of total salinity, the authors focused on the impact of sulfate and calcium ion concentration on oil recovery. The authors obtained the highest oil recovery, in case a sulfate-enriched Sea-water was used as injection fluid. They furthermore concluded that decreasing the calcium ion concentration increases the oil recovery. The most inefficient oil recovery was observed when deionized water was used as secondary injection fluid. All corefloodings were conducted on Middle East carbonate reservoirs samples.

Indiana limestone outcrop cores were used by Shehata et al. [91] to compare the potential of different brine compositions to increase oil recovery in tertiary injection mode. After the injection of Sea-water and bumping up the injection rates, the subsequent injection of deionized water caused a minor oil recovery increase of 2.8 %. However, a second coreflooding demonstrated that also the reversed brine injection pattern caused an oil recovery increase. After the secondary injection of deionized water, the tertiary Sea-water injection resulted in an additional oil recovery of 4.5 %.

While Al-Attar et al. [4] and Shehata et al. [91] obtained high oil recovery in the case of sulfate-rich injection fluids, the majority of the low-salinity limestone studies focused on diluted imbibing brines. After changing the injection brine from Formation-brine to Sea-water, Gupta et al. [45] reported an incremental oil recovery of 5.1 % from a limestone reservoir sample. In comparison, the tertiary injection of a sulfate-free Sea-water into a reference limestone sample led to an incremental oil recovery of 9 %. The overall highest tertiary oil recovery of 15.5 % and 21.3 % was observed for borate

3.3 COREFLOODING EXPERIMENTS

(BO_3^{3-}) and phosphate (PO_4^{3-}) enriched injection brines. None of the conducted corefloodings included rate bumping.

Tetteh et al. [99] combined zeta potential, interfacial tension and coreflooding measurements to investigate the brine composition impact on oil recovery in Indiana limestone outcrop samples. Besides the total salinity, the authors evaluated the calcium, magnesium and sulfate ion concentration impact on oil recovery in tertiary injection mode. After the injection of Formation-water in secondary mode, the tertiary injection of Sea-water resulted in an additional oil recovery of 4.5 %. The subsequent injection of Diluted-sea-water caused an additional average oil recovery of 2.3 %. The authors concluded that the reduction of the total salinity promotes wettability alteration towards stronger water-wet conditions and hence increases oil production. Tetteh et al. [99] furthermore assumed that a reduction of the calcium ion concentration significantly contributes to wettability alteration.

Winoto et al. [106] tested the capability of 20 times diluted Sea-water to increase oil recovery after the primary injection of Sea-water. In the case of limestone outcrop samples, tertiary injected 20 times diluted Sea-water caused an average additional oil recovery of 1 %. A much higher additional oil recovery was observed in case the corefloodings were conducted on carbonate reservoir cores. In tertiary injection mode, 20 times diluted Sea-water caused an average additional oil recovery of 10 %. The work of Winoto et al. [106] did not include rate bumping.

Yousef et al. [108, 109] demonstrated the potential of Diluted-sea-water to enhance oil recovery in carbonates at reservoir conditions. After the injection of Sea-water in secondary mode and bumping up the injection rates, three different versions of Diluted-sea-water were consecutively injected into two Middle East reservoir cores. The tertiary injection mode tests resulted into an additional oil recovery of 7 % and 8.5 %, respectively for 2 times diluted Sea-water, 9 % and 10 %, respectively for 10 times diluted Sea-water and 1.6 % and 1 %, respectively for 100 times diluted Sea-water.

3.3 COREFLOODING EXPERIMENTS

Chandrasekhar and Mohanty [22] combined spontaneous imbibition, USS corefloodings and contact angle measurements to study the brine composition impact on oil recovery on limestone reservoir samples. In secondary injection mode, Formation-water recovered 40 %, Sea-water recovered 47 % and 50 times diluted Sea-water recovered 85 % oil. As a result of the high oil production in secondary mode, Chandrasekhar and Mohanty [22] furthermore tested the efficiency of the 50 times diluted Sea-water to increase the oil production in tertiary mode. After the injection of Formation-water, the injection of the Diluted-sea-water increased the oil recovery from 40 to 72 %. The results are in line with their spontaneous imbibition experiments, in which the Diluted-sea-water caused by far the highest oil recovery. Rate bumping was not conducted.

The study of Nasralla et al. [82] included different coreflooding designs to test the potential of Diluted-sea-water to increase oil production. After the secondary injection of Formation-water at a maximum injection rate of 0.025 and $0.2\,cm^3/min$, respectively, the subsequent tertiary brine injection resulted into an additional oil recovery of approximately 10 % and 2.5 % for Sea-water, followed by an additional oil recovery of approximately 5 % and 1.5 % for 3 times diluted Sea-water, followed by an additional oil recovery of 1 % and 2 % for sulfate-enriched low-salinity brine and an additional oil recovery of 1 % and 2 % for the injection of 25 times diluted Sea-water. However, in case the secondary injection of Formation-water included a rate bumping to overcome capillary end-effect, no additional recovery was observed for tertiary low-salinity water injection. This observation questions the conclusions of many USS corefloodings studies, in which the secondary mode injection excluded rate bumping. Without the conduction of rate bumping, Nasralla et al. [82] indicated, that tertiary recovered oil is produced due to capillary end effect attenuation. In this case, the obtained additional oil recovery is not representative of field case recovery. A detailed discussion is provided in Chapter 7.1 and Chapter 9.

In line with the spontaneous imbibition tests, the available literature on low-salinity corefloodings evaluates the impact of potential determining ions and salinity reduction on low-salinity effects in limestones. While [4, 91, 99] emphasized the impact of sulfate and calcium ion on oil recovery, [22, 45, 82, 106, 108, 109] indicate, that reducing the total ionic strength appears to be the more promising approach to promote low-salinity effects in limestones. However, compared to spontaneous imbibition tests, the conduction and analysis of corefloodings experiments are more complex. Nasralla et al. [82] pointed out, that the injection pattern and design of a coreflooding significantly affect the experimental outcome and hence complicates the comparison of the different studies.

3.4 Low-salinity mechanisms

While the low-salinity research on carbonates initially focused on chalk formations, the literature increasingly covers limestone and dolomite reservoir. The previous sections pointed out that the literature predominately discusses two concepts to promote low-salinity effects in limestones: The modification of the overall potential determining ions and/or the lowering of the total ionic strength [89]. The literature review referred to several spontaneous imbibition experiments, in which highly diluted imbibing brines caused significant primary spontaneous imbibition and additional spontaneous oil recovery in tertiary imbibition mode. In the case of USS coreflooding experiments, the available literature is less consistent. Some studies reported improved oil recovery due to the introduction of PDI into the system, while other studies suggested the reduction of the total ionic strength.

Despite the carbonate composition/genesis and low-salinity approach, it is generally acknowledged that a wettability alteration towards stronger intermediate-wet or water-wet conditions is the main reason for improved oil recovery [9, 14, 64, 66, 96].

Some of the listed coreflooding studies (Section 3.3) included the measurement of the arising oil-water-limestone contact angles. Tetteh et al. [99] measured an approximate average contact angle of 155° for Formation-water, 121° for Sea-water, and 103° for different versions of diluted water. By definition, a contact angle between 0°

3.4 LOW-SALINITY MECHANISMS

and 75° is defined as water-wet, a contact angle between 75° and 115° as intermediate-wet and a contact angle between 115° and 180° as oil-wet [10].

The contact angle measurements of the work of Yousef et al. [109] were conducted at reservoir conditions. In the case of Formation-water and Sea-water, a contact angle of 90° was measured. As a function of brine dilution, the arising contact angle consecutively decreased towards 62.2° for 2 times, 10 times, 20 times and 100 times diluted Sea-water.

In line with Tetteh [99] and Yousef et al. [109], Mahani et al. [64] reported a continues decrease of the contact angle, as the Formation-water was exchanged by Sea-water and Diluted-sea-water, respectively.

Although wettability alteration is a reasonable explanation of improved oil recovery in low-salinity applications, the physical and/or chemical mechanisms behind wettability alteration is still controversial. While Sheng [94] enumerated seventeen proposed low-salinity mechanisms for sandstone and carbonates, Alotaibi et al. [9] and Mahani et al. [64] listed four proposed mechanisms for carbonates: (a) Mineral dissolution/precipitation [51], (b) Surface charge change [111], (c) In-situ surfactant generation [71] and (d) a combination of Mineral dissolution and Surface charge change [110].

Mineral dissolution was initially proposed by Hiorth et al. [51], who concluded that a change in the surface carbonate charge results in mineral dissolution (and consequently oil mobilization). Based on the comparison of the (NMR) T_2 signal before and after low-salinity flooding, Yousef et al. [109] assumed carbonate dissolution as the driving mechanism of low-salinity effects. As a result of pore connectivity increase, Yousef et al. [109] supposed a wettability alteration towards stronger water-wet conditions occurred.

In contrast, Romanuka et al. [89] and Nasralla et al. [82] questioned the impact of mineral dissolution on salinity effects in carbonates [65]. In line with Romanuka et al. [89] and Nasralla et al. [82] this study did not observe any signs of mineral dissolution (cf. Section 4.3.7).

3.4 LOW-SALINITY MECHANISMS

Mahani et al. [64] furthermore excluded the proposed concept of In-situ surfactant as a possible low-salinity mechanism, as the mechanism requires a strongly alkaline environment. This literature review therefore focuses on the concept of Surface charge change.

A possible wettability alteration process based on Surface charge change was sketched by Mahani et al. [66]. The initial conditions are plotted in Figure 3.2, a. Throughout millions of years, an equilibrium between rock surface, oil and reservoir brine establishes. In line with the generally assumed intermediate to oil-wet wettability conditions of carbonates [8, 25, 51], the oil phase tends to attach to the limestone surface. Nevertheless, a thin water film remains at the interface between the rock surface and the oil phase.

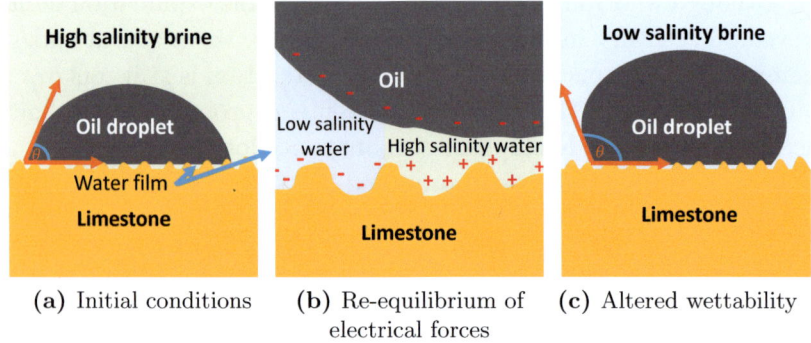

(a) Initial conditions (b) Re-equilibrium of electrical forces (c) Altered wettability

Figure 3.2: Proposed wettability alteration in limestones - The wettability alteration process assumes the low-salinity water diffusion into the water film between the limestone surface and oil phase (a). As a result of the arising Surface charge change, the repulsive oil-limestone forces are strengthened (b). The wettability is changed towards stronger water-wet conditions (c) (redrawn after [66]).

Under the presence of the high saline Formation-water, the initially positively charged limestone surface attracts the negatively charged oil components (Figure 3.2, b, right). In the initial state, a significant amount of the limestone surface is hence coated by oil. However,

3.4 LOW-SALINITY MECHANISMS

as the injected or imbibed lower saline water diffuses into the water film between the oil phase and limestone surface, the limestone surface charge changes. Besides attracting multivalent anions, Surface charge change might be supported by an Electrical double layer expansion and/or mineral dissolution. Figure 3.2, b pictures the re-equilibrium of the oil-brine-solid system, in which the arising repulsive solid-oil forces are causing the detachment of the oil phase. As a result of the oil-brine-solid re-equilibrium, the wettability changes towards stronger water-wet conditions (Figure 3.2, c).

The sketched wettability alteration assumes Surface charge change as the driving mechanism of low-salinity effects. The mechanism is based on experimental observations which indicated, that carbonate Surface charge varies as a function of brine composition, salinity, pH value and temperature [9, 64, 65, 96, 103, 111]. While, for example, the covalent bond of quartz molecules are believed to cause a fixed isoelectric point, the complex chemistry of carbonates allows the modification of the surface charge.

In the first place, it is believed, that a change of the potential determining ions concentration (Ca^{2+}, Mg^{2+}, SO_4^{2-}, CO_3^{2-}) in the vicinity of the carbonate surface causes a re-equilibrium of the electrostatic forces. Furthermore, the overall reduction of the system's salinity can lead to the expansion of the Electrical double layer. Consequently, Surface charge change is the result of the unique or combined appearance of the re-equilibrium of the electrostatic forces and/or double-layer expansion [66, 83]. Several carbonate Surface complexation models are existing in the literature and allow an improved understanding of the proposed Surface charge change mechanism ([21, 103] and more recently [20, 66, 96]).

Zeta potential measurements are the commonly applied method to evaluate surface charges. In general, zeta potential describes the electric potential of a colloid suspension and is hence strongly linked to the concept of Electrical double layer (EDL) [90]. In case the insoluble particle of a colloid exhibits a surface charge, the charge is compensated by the generation of two electrical layers around the particle surface.

3.4 LOW-SALINITY MECHANISMS

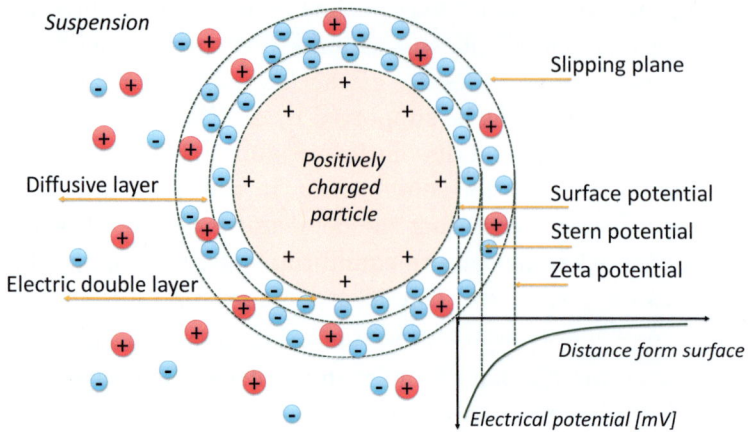

Figure 3.3: Idealized Electrical double layer - Schematic illustration of an Electrical double layer at the surface of a positively charged particle (redrawn after [85]).

An illustration of an idealized Electrical double layer is illustrated in Figure 3.3. Figure 3.3 assumes a positively charged particle surface, which leads to the attraction of anions around the solid surface. The arising first ion layer is referred to as surface charge, which is caused by chemical bonds between the ions and the solid surface. As a result of the surface charge, a second electrical layer is attracted, in which Coulomb forces cause a loosely bond between the ions and surface charge. Compared to the strong chemical bonds of the surface charge layer, the second layer is characterized by a much more diffusive bond between the ions and particle surface.

As the particle and the corresponding double layer are moving through the suspension, friction forces randomly cause the partial separation of the diffusive layer. Consequently, the arising electrical potential at the slipping plane can be measured, which is referred to as zeta potential.

A correlation between brine composition and chalk surface charge was first reported by Zhang et al. [111]. In case of a Ca^{2+} and Mg^{2+} ion concentration increase, the zeta potential measurement

3.4 LOW-SALINITY MECHANISMS

of the tested chalk-brine suspensions increased while the admixture of SO_4^{2-} ions resulted into negative zeta potential values. In line with the spontaneous imbibition and USS coreflooding literature review, the following section focuses on the available literature on zeta potential measurements on limestone-brine systems.

The work of Song et al. [96] includes zeta potential measurements of a sequence of (synthetic) calcite-brine suspensions. Compared to a zeta potential of approximately $+10\,mV$ of a $NaCl$-calcite reference suspension, the zeta potential measurements of a Mg^{2+} enriched brine-calcite and Ca^{2+} enriched brine-calcite suspension resulted into approximately two-times higher zeta potential values. In contrast, the zeta potential of CO_3^{2-} and SO_4^2 enriched calcite-brine suspensions caused strongly negative zeta potential values [96].

A similar trend was demonstrated by Jackson et al. [54]. The zeta potential measurements of the Formation-water and limestone suspension yielded into approximately $+6\,mV$, while the Sea-water and limestone suspension ($\approx$ -$2\,mV$) and the 20 times diluted Sea-water and limestone suspension resulted into negative zeta potential values ($\approx$ -$8\,mV$).

A very comprehensive study on zeta potential measurements on carbonate-brine systems was published by the research group of Mahani et al. [14, 64, 65, 66]. At a pH value of 7 and in line with the findings of Jackson et al. [54], following zeta potential were measured: Formation-water and limestone $\approx +5\,mV$, Sea-water and limestone $\approx$ -$3\,mV$ and 25 times diluted Sea-water limestone $\approx$ -$13\,mV$ [64].

Mahani et al. [65] additionally examined the impact of carbonate type on zeta potential. Figure 3.4 summarizes the obtained zeta potential measurements of dolomite, calcite, limestone, chalk and Formation-water, Sea-water and 25 times diluted Sea-water suspensions. While all four carbonates types confirm a correlation between decreasing salinity and reducing the zeta potential (stronger negative), the Surface charge change sensitivity differs for each material. In general, dolomite exhibited the most positive zeta potential while, for example, chalk exhibited a negative zeta potential in all tested brine-carbonate suspensions. The plotted zeta potential measure-

3.5 SUMMARY & CONCLUSIONS

ments furthermore indicate an explanation of the varying low-salinity response of chalk and limestone samples. While the low-salinity effects of chalks samples are predominately linked to the PDI concentration, the total ionic strength significantly impacts the surface charge of limestone samples.

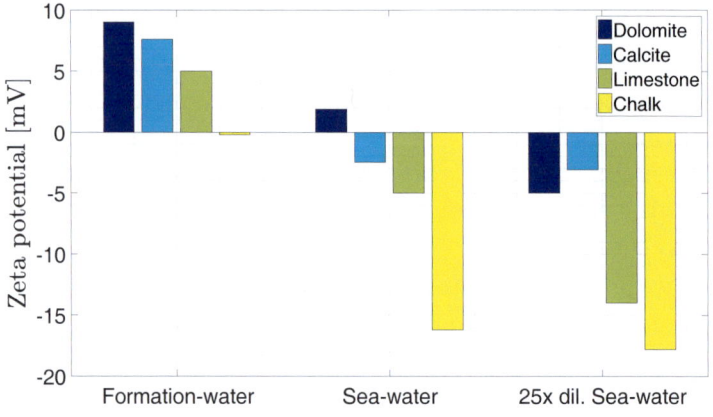

Figure 3.4: Carbonate type impact on zeta potential - Zeta potential measurement results of dolomite, calcite, limestone, chalk and Formation-water, Sea-water and 25 times diluted Sea-water suspensions (redrawn after [65]).

3.5 Summary & conclusions

Based on the Buckley-Leverett solution, the literature review initially illustrated the secondary and tertiary low-salinity waterflooding. The reviewed limestone spontaneous imbibition tests demonstrated the potential of highly diluted imbibing to promote spontaneous oil recovery. Compared to the injection of Formation-water, the unsteady state experiments showed improved/additional oil production for PDI enriched as well as diluted injection waters. However, the majority of the published work did not include rate bumping, which questions the recorded additional oil recovery.

3.5 SUMMARY & CONCLUSIONS

The research group of Mahani et al. [14, 64, 65, 66] proposed Surface charge change as the driving mechanism of low-salinity effects in carbonates. The proposed mechanism explains the varying response of dolomite, limestone and chalks on low-salinity applications. Based on the suggested Surface charge change mechanism, this study combines zeta potential, spontaneous imbibition, centrifuge method and coreflooding experiments to evaluate the physics of the observed oil recovery.

Part I
Experimental study

4

Preparation

A careful and unified sample preparation is crucial for the success of any experimental study. Divided into three subsections, this chapter describes the fluid preparation, core sample treatment and the establishment of the connate water saturation.

4.1 Fluid preparation

4.1.1 Brine preparation

The experimental study uses three different brines to investigate the brine composition impact on oil recovery: A high-saline Formation-water (salinity of $183.4\,g/l$), Sea-water ($43.8\,g/l$) and (100 times) Diluted-sea-water ($0.4\,g/l$). The three brines were synthesized based on a provided ionic composition analysis. Since brine synthesis includes the admixture of salt into deionized water, the ion concentration of the ionic composition analysis was initially converted into salt fractions

$$n = \frac{m}{M}, \quad (4.1)$$

where n is the amount of substance in $[kmol]$, m is the mass in $[kg]$ and M is the molar mass in $[kg/kmol]$. The brine synthesis

4.1 FLUID PREPARATION

Brine	Sodium [mol]	Potassium [mol]	Calcium [mol]	Magnesium [mol]	Strontium [mol]	Chlorine [mol]	Bromine [mol]	Sulfate [mol]	Salinity [g/l]	Ionic strength
Formation water	2.237	0.023	0.372	0.084	0.009	3.165	0.009	0.004	183.4	3.655
Sea water	0.59	0.01	0.01	0.07	0.0001	0.68	0	0.04	43.8	0.870
100 times Diluted-SW	0.0059	0.0001	0.0001	0.0007	0.000001	0.0068	0	0.0004	0.438	0.0087

Table 4.1: Brine composition - Brine composition of Formation-water, Sea-water and (100 times) Diluted-sea-water.

required the admixture of overall seven salt compounds: Sodium chloride ($NaCl$), sodium sulfate (Na_2SO_4), calcium chloride dihydrate ($CaCl_2 \cdot 2H_2O$), magnesium chloride hexahydrate ($MgCl_2 \cdot 6H_2O$), potassium chloride (KCl), strontium chloride hexahydrate ($SrCl_2 \cdot 6H_2O$) and potassium bromide (KBr).

While the initial ionic composition analysis included traces of bicarbonate, the admixture of the organic sodium bicarbonate ($NaHCO_3$) caused the generation of a cloudy brine suspension. Consequently, sodium bicarbonate was excluded from the synthesis. Furthermore, the low solubility of sodium sulfate was overcome by initially dissolving the salt inside a separated solution.

Before using the brines in the experimental study, the prepared solutions were filtrated to remove solids and other impurities. The brine compositions of the Formation-water, Sea-water and Diluted-sea-water are summarized in Table 4.1.

4.1.2 Oil preparation

A Middle East light dead crude oil was used in the study. In general, waxes and other heavy oil components tend to complicate experimental work. For instance, heavy oil components distorted the optical in-situ saturation monitoring of the forced drainage and imbibition tests. In order to reduce the number of oil waxes, the crude oil was therefore centrifuged for $30\,min$ at a centrifuge spin of $5000\,RPM$.

4.1 FLUID PREPARATION

In comparison to the more commonly applied oil filtration, the oil centrifugation caused a much more efficient heavy oil component separation. An optical impression of the separated high viscous oil waxes is displayed in Figure 4.2, a.

4.1.3 Density & viscosity

An Anton-Paar densitometer and viscometer was used to measure the oil density and viscosity within a temperature range of 20 to 50°C. As the application of higher temperatures can cause the vaporization of oil components, oil viscosity and density interpolations to temperatures above 50°C were required. By convention, the fluid property measurements were started at the maximum test temperature of 50°C to ensure the heavy oil component dissolution.

The high salinity of the Formation-water and Sea-water tends to corrode the measuring cells of many commercial densitometers and viscometers. While the brine density was measured with the help of a specialized Anton-Paar densitometer, the brine viscosity was therefore determined by a conventional capillary viscometer. The brine properties were measured within a temperature range of 30 to 60°C.

In regard to the spontaneous imbibition, forced imbibition and unsteady state displacement experiments, the obtained density and viscosity measurements required an interpolation to 70°C. While the temperature impact on density can be described by a simple linear function, the non-linear temperature/viscosity relation requires an application of a specialized mathematical description (cf. Figure 4.1). This study uses the Arrhenius equation to interpolate the temperature impact on viscosity [75]

$$\ln(\mu) = \ln(A_s) + \frac{E_\alpha}{R}\left(\frac{1}{T}\right), \quad (4.2)$$

where μ is the viscosity in $[Pas]$, A_s is the pre-exponential factor, E_α is the activation energy in $[J/mol]$, R is the (universal) gas constant in $[J/molK]$ and T is the temperature in K.

4.1 FLUID PREPARATION

An example of the Arrhenius equation is provided in Table 4.2, in which the viscosity of the Formation-water is interpolated. Equation 4.2 is solved by plotting the reciprocal (Kelvin) temperature versus the logarithm of the experimentally obtained viscosity values. In case the graph yields into a linear correlation, the activation energy E_α and the pre-exponential factor A_s are temperature independent. The temperature independence allows some fundamental simplifications. First, the activation energy E_α/R is determined by the slope of the reciprocal temperature and the logarithmic viscosity function. In the second place, the y-intercept of the linear reciprocal temperature/ logarithmic viscosity function corresponds to the pre-exponential factor $\ln(A_s)$. As a result of the geometrical E_α/R and $\ln(A_s)$ calculation, the interpolated viscosity is obtained by inserting the reciprocal temperature into Equation 4.2. The measured and interpolated density and viscosity values are displayed in Figure 4.1.

Temp. [°C]	Meas. μ [mPas]	1/Temp. [K]	$\ln(\mu)$			Arrhenius μ [mPas]
70						0.625
60	0.728	0.00300	-0.3174	$\mathbf{E_\alpha/R}$		0.723
50	0.839	0.00309	-0.1760		1672.5	0.845
40	0.993	0.00319	-0.0073	$\mathbf{\ln(A_s)}$		0.997
30	1.195	0.00330	0.1778		-5.34	1.189
20						1.435

Table 4.2: **Arrhenius viscosity interpolation** - The viscosity interpolation to 70°C is based on the application of the Arrhenius equation.

4.1.4 Interfacial tension

Interfacial tensions (IFT) are generated by the imbalance of the molecular forces at the surface of two immiscible phases [15]. As a result of interfacial tensions, immiscible fluids are prevented from dissolving into each other. In regard to conventional waterfloodings,

4.1 FLUID PREPARATION

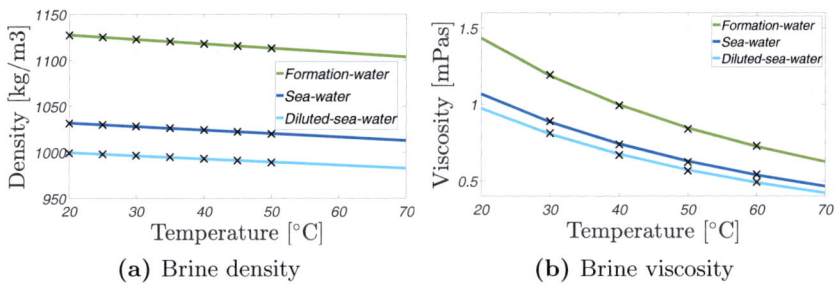

(a) Brine density

(b) Brine viscosity

Figure 4.1: Brine density and viscosity - The temperature/density correlation is linear (a), while the non-linear temperature/viscosity correlation is displayed in (b).

the interfacial tension between the immiscible oil and water phases significantly impacts oil recovery. In general, high interfacial tensions decrease the displacement efficiency, while ultra-low IFT can cause a residual oil saturation reduction [104].

The IFT between the crude oil and three different brines were measured by applying the Pendant-drop method. All measurements were conducted at room temperature and ambient pressure conditions. Thereby, a small syringe was initially filled with crude oil and then lowered into a transparent brine filled container. After forming an oil bubble at the end of the needle, the arsing bubble shape was captured by a camera. The buoyancy force was considered by bending the syringe needle by approximately 180°. Once a static and symmetric drop was obtained, the corresponding interfacial tension was calculated based on the Young-Laplace equation [72]

$$\sigma = \frac{g \cdot \rho \cdot d_e^2}{H_{PD}}, \quad (4.3)$$

where σ is the surface tension in $[N/m]$, g is the gravity in $[m/s^2]$, ρ is the droplet density in kg/m^3 and H_{PD} is a geometrical correction factor. An example of a Pendant-drop measurement is pictured in Figure 4.2, b where the correction factor H_{PD} depends on the

4.2 CORE PREPARATION

maximum and minimal droplet diameter d_e/d_s. At ambient conditions, following interfacial tension were measured: Formation-water and crude oil 27.8 mN/m, Sea-water and crude oil 27.3 mN/m and Diluted-sea-water and crude oil 24.8 mN/m.

(a) Crude oil separation

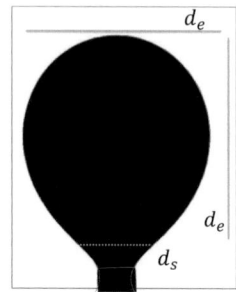
(b) Pendant-drop method

Figure 4.2: Crude oil properties - The separated crude oil waxes are displayed in (a), while (b) illustrates an example of a Pendant-drop measurement.

4.2 Core preparation

4.2.1 Core material

The experimental study was conducted on Indiana limestone outcrop samples. The samples were purchased from Kocurek Industries, which extracted the samples from a limestone outcrop formation in Bedford, Indiana, USA. The literature reports several studies in which the Kocurek Industries Indiana limestone samples were used. Freiere et al. [41] conducted an X-ray fluorescence composition analysis on the limestone samples, which revealed calcite as the major component (98.6 wt %). Further components are magnesium oxide (0.56 wt %) and quartz (0.35 wt %). The complete mineral composition analysis of the Indiana limestone samples is listed in Table 4.3.

4.2 CORE PREPARATION

Compound	$CaCO_3$	MgO	SiO_2	FeO	Al_2O_3	K_2O	Cl	S	Na_2O	Sr	P_2O_5
Atomic weight [%]	98.58	0.56	0.35	0.14	0.12	0.08	0.06	0.05	0.03	0.02	0.01

Table 4.3: **Indiana limestone composition** - X-ray fluorescence composition analysis of Indiana limestone samples [41].

4.2.2 Core cleaning

During the process of reservoir wettability restoration/establishment, core cleaning represents a key challenge in which contaminations such as mud-filtrates, oil, water and/or evaporated salts are removed from the samples [59, 72]. Usually, the cores are exposed to a non-polar (e.g., toluene and chloroform) and a polar (e.g., acetone and methanol) solvent. While non-polar solvents are dissolving light crude oil components, polar solvents are used to remove heavier oil components and/or precipitated salts [72]. Since the Indiana limestone samples were extracted from an outcrop, the cleaning procedure was limited to the application of a methanol solvent.

All samples were cleaned inside an automatized flush-through cleaning rig. The set-up consists of several parallel aligned core holders, which ensures the simultaneous cleaning of up to six cores sample. The injection of the cleaning solvent is thereby monitored by a pump control system, which automatically adapts the injection rates to the obtained differential pressure.

Compared to the more commonly applied Soxhlet-extraction cleaning procedure, the flush-through cleaning system has several advantages. Besides the efficiency of the method, the cleaning progress of the flush-through system can be optically estimated by monitoring the effluent color. In the case of the Indiana limestone samples, the effluent color changed from yellow to colorless after the injection of approximately 10 pore volumes of methanol. The cleaning process was completed by a subsequent vacuum drying of 48 *hours* at 90°C.

4.2 CORE PREPARATION

4.2.3 Core selection

In regard to the centrifuge dimensions, the $3.7\,cm$ ($1.5\,inch$) thick limestone samples were trimmed and smoothed down to $5\,cm$ ($2\,inch$) lengths. The samples were then imaged by a Medical computed tomography scan to identify and sort out fractured and damaged cores. A caliper was used to measure the sample's core length and diameter.

The implemented study includes eight different connate and imbibing water combinations. Each brine combination was tested on two core samples: Preferably a less permeable and a higher permeable limestone sample. To ensure an appropriate sample allocation, helium porosity and absolute nitrogen permeability were measured. The corresponding core selection is plotted in Figure 4.3.

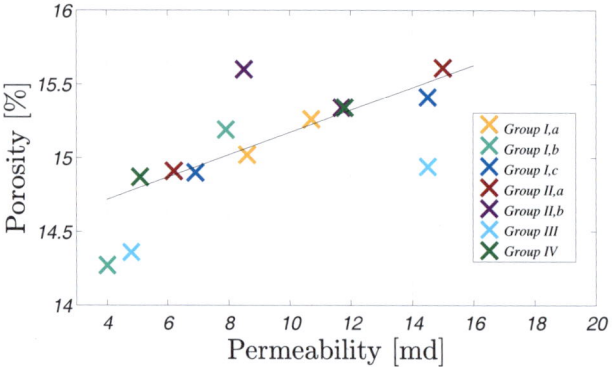

Figure 4.3: Core selection - The core selection is based on absolute permeability. Each brine configuration was preferably tested on a less permeable and a higher permeable sample.

4.3 Connate water saturation

4.3.1 Core saturation

After cleaning and drying the cores, the dry weights of the core samples were recorded. The cores were then placed inside a steal accumulator to saturate the samples with the connate water. In the first place, the accumulator was evacuated for 3 $hours$ to remove air from the core samples. The cores samples were subsequently pressurized by the selected connate water for 48 $hours$ at 138 bar (2000 psi). After removing the cores from the saturation equipment, the core wet weights were measured. The sample porosity was determined based on Equation 4.4

$$V_P = \frac{m_{wet} - m_{dry}}{\rho_i} \times \frac{1}{V_C}, \qquad (4.4)$$

where V_P is the pore volume in $[m^3]$, m_{wet} is the wet weight in $[kg]$, m_{dry} is the dry weight in $[kg]$, ρ_i is the density of connate water in $[kg/m^3]$ and V_C is the core sample volume in $[m^3]$.

4.3.2 Absolute brine permeability

The study included the absolute brine and effective oil permeability measurements before and after aging. Generally, absolute permeability measurements are conducted at a fluid saturation of 100 %, while the effective permeability refers to permeability measurements under the presence of at least two immiscible fluids [72]. In regard to the overall 51 permeability measurements, a simple and efficient permeameter was developed.

The set-up of the absolute brine permeability measurement is sketched in Figure 4.4, a. To promote a homogeneous and steady fluid flow, a vertical core alignment was selected over a horizontal alignment. The core samples were initially mounted inside the core sleeve and then pressurized by a confining pressure of 35 bar (500 psi). In the case of the absolute permeability measurements,

4.3 CONNATE WATER SATURATION

the brine was injected through the bottom face of the core holder. After starting the brine injection and obtaining the first effluent production, a manual pressure regulator was used for the step-wise back-pressure establishment of $7\,bar$ $(100\,psi)$. The pressure drop along the core was determined by the pressure difference between the inlet and outlet face pressure transducers.

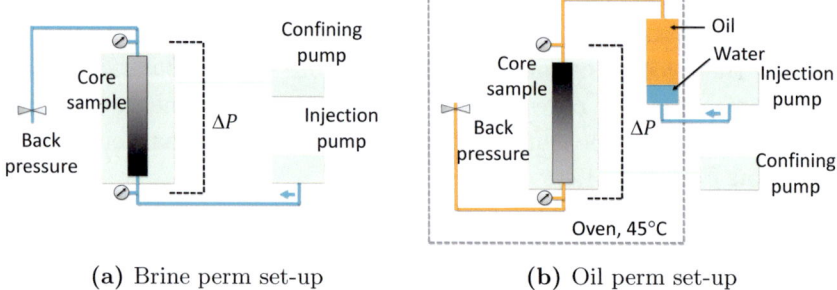

(a) Brine perm set-up   (b) Oil perm set-up

Figure 4.4: Absolute brine and effective oil permeability measurement set-up - To promote a gravity stable fluid flow, the injection face was adapted to the respective measurement.

In comparison to the effective oil permeability measurements, the absolute permeability experiments were less failure-prone and time-consuming. However, a unified data analysis procedure helped to generate and identify reliable permeability results. In the first place, the linear correlation between the applied injection rates and the corresponding differential pressure values was calculated using Pearson's correlation coefficient [12]

$$Pearson = \frac{\sum_{i=1}^{n}(q_i - \bar{q})(\Delta P_i - \Delta \bar{P})}{\sqrt{\sum_{i=1}^{n}(q_i - \bar{q})^2 \sum_{i=1}^{n}(\Delta \bar{P}_i - \Delta \bar{P})^2}}, \quad (4.5)$$

where q_i denotes the applied injection rates, $\bar{q}$ is the mean injection rate, ΔP_i denotes the differential pressures and $\Delta \bar{P}$ is the mean differential pressure. The results of the absolute and effective permeability measurements are attached in Appendix A. While a Pearson's

4.3 CONNATE WATER SATURATION

correlation coefficient of 1 expresses a perfect correlation, the vast majority of the absolute brine permeability measurements resulted in a Pearson's correlation coefficient of 0.9996 and 1.

Besides Pearson's correlation coefficient as the first indicator of a successful measurement, a differential pressure offset calculation was conducted to revise the obtained pressure data. During permeability measurements, several factors such as insufficient pressure transducer calibration, temperature effects, atmospheric pressure changes or improper sample mounting can cause a distortion of the obtained pressure values. Whereas the origin of these factors can hardly be avoided during laboratory work, the quantitative value of the distortion (offset) can be geometrically determined. After plotting the applied injection rate versus the corresponding differential pressure, the y-intercept represents the quantity of the pressure offset. The obtained differential pressure values at each injection rate are corrected by the subtraction of the offset pressure value. In regard to the temperature impact of the effective oil permeability measurements, the absolute brine permeability measurements were characterized by smaller pressure offset values.

Besides the correct sample mounting, experimental conduction and data correction of the permeability measurements, the applied injection rates were furthermore adapted to the pressure sensibility of the pressure transducers. In general, high permeable samples require the application of higher injection rates to obtain a reliable minimum differential pressure.

After calculating Pearson's correlation coefficient and correcting the differential pressure values by the consideration of the pressure offset, the absolute brine permeability was calculated based on Darcy's Law [2]

$$k = \frac{q_{lab} \cdot \mu \cdot L}{A \cdot \Delta P}, \tag{4.6}$$

where k is the permeability in $[m^2]$, q_{lab} is the volumetric injection rate in $[m^3/s]$, μ is the viscosity in $[Pas]$, L is the core length in $[m]$,

4.3 CONNATE WATER SATURATION

A is the core cross-section in m^2 and ΔP is the differential pressure in $[bar]$. An example of a standardized permeability measurement and analysis is provided in Table 4.4.

		Flowrate $[cm^3/min]$	$\Delta \rho$ $[barg]$	Corr. $\Delta \rho$ $[barg]$	Perm. $[md]$
Sample	In10	0.2	0.352	0.334	6.32
Length [mm]	37.94	0.5	0.882	0.864	6.10
Diameter [mm]	48.69	1	1.733	1.715	6.15
Viscosity [mPas]	1.45	1.5	2.579	2.561	6.18
Pearson's coef.	**0.9999**	Offset	**0.018**		$\bar{k} = \mathbf{6.19}$

Table 4.4: Standardized permeability measurement example - The permeability measurement analysis included the calculation of Pearson's correlation coefficient as well as the pressure offset.

4.3.3 Nuclear Magnetic Resonance

Nuclear magnetic resonance (NMR) tests are a frequently used method to characterize the pore structure of core samples. The study used a Magritek 2 MHz NMR rock core analyzer to obtain pore-size distributions and connate water saturations.

The fully brine saturated cores were initially wrapped inside clingfilm and then placed inside the NMR device. At the beginning of the measurement, the water protons are parallel aligned to the substantial magnetic field. Once the NMR device starts to generate an oscillating magnetic field, the water protons are forced to change their orientation. While the oscillating magnetic field is applied, the NMR equipment measures the signal amplitude and decay of the energized protons.

The amplitude of the signal decay (also referred to as relaxation time) is impacted by several factors. Besides the molecular motion and diffusion of the brine, especially the collision of the energized water protons and the pore surface is causing a quick signal relaxation.

4.3 CONNATE WATER SATURATION

Since the hydrogen proton and pore surface collision probability increases with decreasing pore-size, short relaxation times are indicative of smaller pore structures [72].

The T_2-relaxation times of the limestone samples are summarized in Figure 4.5. The two amplitude signal peaks indicate double-porosity systems.

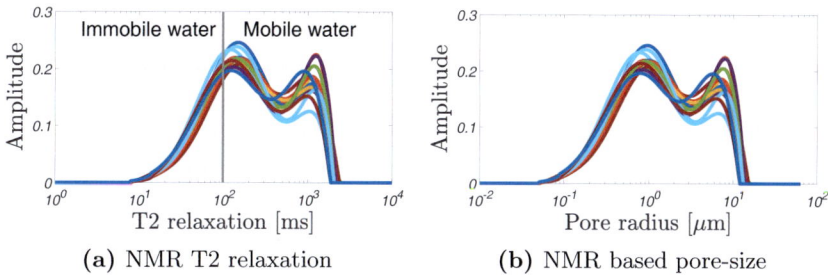

(a) NMR T2 relaxation

(b) NMR based pore-size

Figure 4.5: NMR core characterization - The cut-off time divides the T_2-signal curve into immobile and mobile water volume (left). The pore radius ranges from 0.1 to 10 μm (right).

Based on a material depending cut-off time, the T_2-relaxation time can be furthermore used to derive NMR based Sw_c reference values [72]. Depending on the defined cut-off time, the obtained T_2-relaxation time curve is thereby divided into the Irreducible water Bulk Volume (BVI) and the Free-Fluid Bulk volume (FFV) [26]. The area of the BVI represents the connate or irreducible water, which is immobile due to high capillary forces. After dividing the T_2 relaxation curve into BVI and FFV, the NMR based Sw_c value is calculated by the ratio of the area of the BVI and the total area of the T_2 signal curve. While Chang et al. [23] proposed a carbonate cut-off time of 92 μs, this study found a good connection between the NMR based and experimental obtained Sw_c when a cut-off value of 98 μs was used. The NMR based Sw_c method is illustrated in Figure 4.5, a, where the vertical line separates the immobile and mobile water. Furthermore, a comparison of the NMR based and centrifuge based Sw_c values is provided in Table 4.5.

4.3 CONNATE WATER SATURATION

Similar to the calculation of an NMR based Sw_c reference value, the definition of material depending surface relaxivity value can be used to convert the T_2-relaxation time into pore-size [67]

$$r_p = \frac{T_2 \cdot 2 \cdot \delta}{1000}, \qquad (4.7)$$

where r_p is the pore radius in [μm], T_2 is the relaxation time in [μs] and δ is the surface relaxivity in [$\mu m/s$]. Figure 4.5, b summarizes the results of the pore radius calculation based on a surface relaxivity of 3.2 μs as suggested by Marschall et al. [67]. Based on the proposed surface relativity of 3.2 μs, the limestone pore-size varies between 0.1 and 10 μm.

4.3.4 Primary brine drainage

The establishment of the connate water saturation Sw_c is a crucial prerequisite of any successful oil recovery study. Commonly applied techniques are the coreflooding method, porous plate method and centrifuge technique. Each method has advantages and disadvantages [72]. In regard to the time efficiency and forced imbibition experiments, this study uses the centrifuge method to drain the sample cores to Sw_c. A comprehensive explanation of the centrifuge technique is provided in Chapter 6 and Chapter 8.

The irreducible or connate water saturation is impacted by several factors such as the system's wettability, porosity and permeability. In general, the connate water saturation decreases with decreasing porosity and/or permeability. Furthermore, oil-wet samples usually exhibit smaller Sw_c values than water-wet samples [16]. An insufficient implemented core draining procedure might result in unrepresentative high Sw_c values, which distort the result of Special core analysis experiments. For instance, a mobile water phase (due to insufficient core draining) results in inaccurate capillary pressure measurements (centrifuge method) or promotes an early/immediate water breakthrough during unsteady state displacement experiments.

The difficulty of realizing an appropriate centrifuge drainage run is represented by the definition of the maximum centrifuge velocity.

4.3 CONNATE WATER SATURATION

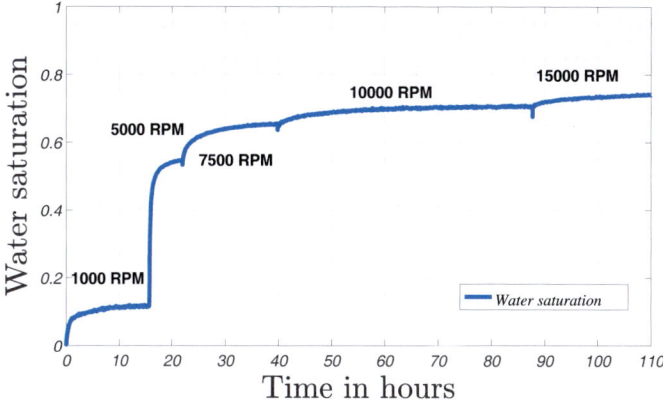

Figure 4.6: **Primary drainage centrifuge method** - In order to define an appropriate maximum centrifuge drainage speed, sample In8 was used to examine the impact of centrifuge spin velocity on water saturation development.

While a low maximum centrifuge velocity results in an unrepresentative high connate water saturation, the emerging pressure of high centrifuge spins can cause sample damage. To define an appropriate centrifuge spin, the primary drainage run of sample In8 was used to investigate the applied centrifuge spin against average water saturation development (Figure 4.6). At a centrifuge speed of 10000 RPM and the corresponding inlet pressure of 10.9 bar, an average water saturation of 70.6 % was obtained. After increasing the centrifuge speed to 15000 RPM, an additional water saturation decrease of 3.6 % was recorded. However, due to the corresponding high inlet pressure of 24.6 bar, the core fractured and was removed from the experimental study. Based on the experience of the test run In8, a single-speed centrifuge drainage rate of 12000 RPM (corresponding inlet pressure of 15.7 bar) was defined. The obtained connate water saturations ranged between 23.0 to 32.5 %.

At first glance, the obtained connate water saturations of 23.0 to 32.5 % might appear unrepresentative high. However, the comparison of the NMR based connate water saturation and physically

4.3 CONNATE WATER SATURATION

determined Sw_c are in good agreement (Table 4.5). It can be concluded that the relativity high connate water saturations are likely caused by the pore geometry and wettability conditions rather than an inappropriate drainage procedure.

4.3.5 Effective oil permeability

In order to ensure the effective oil permeability measurements, the absolute brine permeability measurement set-up of Section 4.3.2 was initially modified. The principle of the effective oil permeability measurement set-up is drawn in Figure 4.4, b. An oil accumulator was assembled between the pump and core holder to avoid pump pollution. After starting the water-filled injection pump, the water entered the oil-filled accumulator through the bottom connection and pushed the oil towards the outlet. In order to attenuate the effect of gravity overriding and channeling, the oil was injected from the top face of the core holder. Since an irregular piston motion tends to distort differential pressure measurements, the accumulator piston was removed from the set-up. In this case, the oil and water phase are separated due to phase immiscibility and gravity force.

During the test runs, the effective oil permeability was initially measured without applying any temperature. However, heavy oil components accumulated at the core injection face and caused an unrepresentative high differential pressure. The effective oil measurements were therefore conducted at 45°C to avoid phase plugging due to the crude oil waxes. In accordance with the absolute brine permeability measurements, a confining pressure of $35\,bar$ and a back-pressure of $7\,bar$ were applied. The results of the effective oil permeability measurements are listed in Table 4.5.

4.3.6 Core aging

The wettability of an oil field is a key factor that impacts the fluid flow, fluid distribution and hence oil recovery of a reservoir [80]. Since carbonates reservoirs are generally believed to behave intermediate to oil-wet [8, 25, 51], the selected limestone outcrop sam-

4.3 CONNATE WATER SATURATION

ples should preferentially be characterized by comparable wettability properties. The commonly applied method to establish oil-wet conditions is referred to as core aging.

During the establishment of the connate water saturation, the samples are initially cleaned, dried and saturated by connate water. It is evident that at this point, the pore surface is entirely coated by the water phase. Due to the subsequent primary drainage, the majority of water is displaced from the core. However, some of the water remains attached to the pore surface. This mainly occurs inside smaller pores, where capillary forces are stronger. As a result of the drainage process, a new equilibrium between the oil, water and limestone surface establishes, which can lead to an additional water displacement from the limestone surface.

To accelerate the oil-water-solid equilibrium establishment, the samples were aged for $30\,days$ at $90\,°C$. Thereby, the core plugs were placed inside a steel accumulator at Sw_c, separated by small plastic meshes and surrounded by crude oil. The accumulator was then placed inside an oven at 90°C and connected to a pump to ensure a constant crude oil pressure of $7\,bar$ $(100\,psi)$.

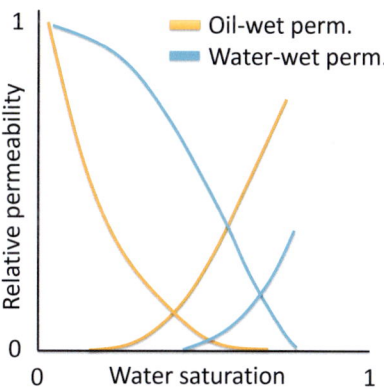

Figure 4.7: Relative permeability comparison - Comparison of a water-wet and oil-wet relative permeability (redrawn after [11]).

4.4 SUMMARY & CONCLUSIONS

Figure 4.7 illustrates a typical water-wet and an oil-wet relative permeability. Based on the curve shapes, a wettability alteration towards stronger oil-wetting conditions is indicated by a reduction of the effective oil permeability at Sw_c. The results of the effective oil permeability before and after aging are summarized in Table 4.5. All samples exhibited a permeability reduction in the range of 9 to 40 %, which indicates a wettability modification towards stronger oil-wetting properties. The oil permeability measurements did not cause additional water displacement. Furthermore, a change in the absolute permeability before and after completing the experimental study was not observed (cf. Section 4.3.7). Although dynamic aging processes are generally believed to be more efficient [70], the conducted static aging process caused a consistent effective oil permeability reduction. All conducted effective oil permeability measurements are listed in Appendix A.

4.3.7 Unsteady state coreflooding samples

After completing the spontaneous and forced imbibition tests, the Indiana limestone samples were cleaned by the injection of water, methanol and toluene. The subsequent core saturation, drainage and aging was conducted in accordance with Section 4.2.2 to Section 4.3.6. Three samples were selected for unsteady state displacement experiments. Table 4.5 shows that the measured properties of In2/In2b, In4/In4b and In9/In9b are almost identical. It can therefore be concluded, that the applied core preparation procedure caused reproducible sample conditions. Furthermore, the spontaneous and forced imbibition tests did not change the sample properties. Dissolution of the calcite surface due to the injection of Formation-water, Sea-water or Diluted-sea-water did not occur.

4.4 Summary & conclusions

The core preparation, fluid preparation and particularly the establishment of the connate water saturation is a crucial process during oil recovery studies. Seventeen core samples were successfully prepared during the study.

4.4 SUMMARY & CONCLUSIONS

Group	Sample	CW	IW	Porosity [%]	Brine perm [md]	Oil perm. before aging [md]	Oil perm.after aging [md]	NMR cut-off [%]	S_{wc} [%]
I	In3	FB	FB	15.0	8.6	9.1	7.3	28.2	27.0
	In5	FB	FB	15.3	10.7	12.6	7.9	28.2	24.5
	In7	SW	SW	14.3	4.0	3.4	2.6	34.9	32.4
	In12	SW	SW	15.2	7.9	7.1	4.9	28.8	32.9
	In4	DSW	DSW	14.9	6.9	5.5	3.8	28.5	31.2
	In13	DSW	DSW	15.4	14.5	9.9	7.0	27.8	29.8
II	In2	FB	SW	15.6	15.0	12.8	11.1	26.2	26.4
	In10	FB	SW	14.9	6.2	5.6	2.5	26.7	30.4
	In1	FB	DSW	15.6	8.5	9.4	7.1	28.4	30.2
	In9	FB	DSW	15.3	11.7	11.5	8.1	23.0	25.7
III	In16	SW	DSW	14.4	4.8	6.4	5.6	32.5	28.6
	In17	SW	DSW	14.9	14.5	11.2	10.2	29.9	25.8
IV	In14	DSW	FB	15.3	11.8	11.2	8.9	26.8	24.0
	In15	DSW	SW	14.9	5.1	5.5	4.9	29.0	26.3
USS	In2b	FB	FB	15.4	15.0	11.9	9.9	-	21.8
	In4b	FB	SW	14.5	7.4	7.2	4.6	-	26.5
	In9b	FB	DSW	15.2	12.3	9.7	7.0	-	24.9

Table 4.5: **Core properties** - A summary of the core properties of the experimental spontaneous and forced imbibition part (top) and the experimental unsteady state coreflooding part (bottom).

- The oil and brine density and viscosity properties were measured in a temperature range of 20 to 50°C and 30 to 60°C,

4.4 SUMMARY & CONCLUSIONS

respectively. A linear equation interpolated the density properties to 70°C, while the Arrhenius equation was used to interpolate the fluid viscosity to 70°C.

- The interfacial tension between the Formation-water and crude oil, Sea-water and crude oil and Diluted-sea-water and crude oil were measured based on the Pendant-drop method. The IFT differences of the three combinations were marginal.

- The study used Indiana limestone samples to conduct spontaneous imbibition, forced imbibition and unsteady state corefloodings experiments. In regard to the centrifuge method, the core length was limited to $5\,cm$. Calcite represented the major sample component ($98.6\,wt\,\%$).

- The cores were cleaned by applying the flush-through cleaning method. After saturating and draining the cores to Sw_c, the cores were aged for $30\,days$. The obtained Sw_c ranged between 23.0 to 32.5 %, which is in line with the NMR based reference values. The effective permeability reduction before and after aging indicates a wettability alteration towards stronger oil-wet conditions.

- A simple and unified absolute brine permeability/effective oil permeability set-up and data analysis were developed. All measurements showed a linear correlation between the applied injection rates and corresponding differential pressures. The results of the absolute and effective measurements are listed in Appendix A.

- After completing the spontaneous and forced imbibition experiments, the limestone samples were cleaned, saturated, drained and aged for the subsequent unsteady state experiments. The reproduction of the initial absolute and effective permeability values indicates a suitable and successful core preparation procedure. The imbibition of the Formation-water, Sea-water and/or Diluted-sea-water did not change the core properties/structures.

5
Spontaneous imbibition

Spontaneous imbibition tests are a frequently used method to screen the spontaneous imbibition behavior of a wide range of imbibing fluids. This chapter describes the conduction of fourteen imbibition tests and links the obtained results to contact angle and zeta potential measurements.

5.1 Theory

The low-salinity literature quickly links spontaneous imbibition tests to complex processes such as wettability alteration and/or possible Enhanced oil recovery effects. During several decades of low-salinity research, a large number of low-salinity mechanisms have been proposed [94]. In regard to limestone carbonates, Chapter 3 provides a comprehensive overview of the proposed mechanisms. However, before considering possible low-salinity mechanisms, this section initially pictures the fundamental physics of spontaneous imbibition.

From a physical point of view, spontaneous imbibition describes the process, in which a wetting fluid imbibes a porous medium due to capillary forces [78]. Spontaneous imbibition can be observed in daily life, for example, when a sugar cube is immersed in tee and is immediately soaked by the imbibing fluid.

5.1 THEORY

An illustration of an idealized water-oil imbibition process into an elongated pore is sketched in Figure 5.1. The imbibing water represents the wetting fluid, which hence tends to attach to the pore surface. As a result of the water-wetness, the arising contact angle θ between the water, oil and solid surface is smaller than 90°. In case the pore geometry allows the concave oil-water interface to advance, the water spontaneously imbibes into the sample [78].

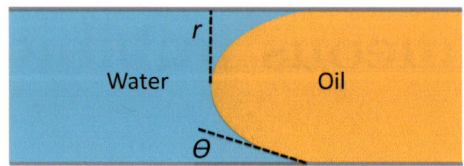

Figure 5.1: Idealized spontaneous imbibition illustration - As a result of the concave water-oil interface, the water spontaneously imbibes into the tube model.

According to its universal definition, spontaneous imbibition is closely linked to capillary pressure. Capillary pressure itself is caused by the interfacial tensions of two immiscible phases, which occur along the phase interface. The quantity of the differential phase pressure P_c can be described by Equation 5.1 [101]

$$P_c = \frac{2 \cdot \sigma_{ow} \cdot \cos(\theta)}{r_p}, \tag{5.1}$$

where P_c is capillary pressure in $[Pa]$, σ_{ow} is the interfacial tension between the oil and water in $[J/m^2]$, θ is the contact angle between the water, oil and solid surface and r_p is the pore radius in $[m]$. Based on Equation 5.1, spontaneous imbibition predominately depends on the interfacial tension σ_{ow} and the solid-liquid-liquid contact angle θ.

The Pendant-drop oil-brine measurements of the Formation-water and oil ($27.8\,mN/m$), Sea-water and oil ($27.3\,mN/m$) and Diluted-sea-water and oil ($24.8\,mN/m$) (cf. Chapter 4.1.4) resulted in negligible small interfacial tension differences. The oil-water IFT impact

5.2 SPONTANEOUS IMBIBITION IN CARBONATES

on the conducted spontaneous imbibition tests is therefore assumed to be independent of the selected brine compositions.

The contact angle impact on the spontaneous imbibition behavior into Polytetrafluoroethylene (PTFE) cores was extensively investigated in the work of Morrow [79]. In case of a wetting contact angle of 0 to 62°, the wetting fluid tended to imbibe into the porous media spontaneously. Thereby, the imbibition rate decreased with increasing wetting fluid saturation. For intermediate wetting contact angles (62 - 133°), neither the wetting nor the non-wetting spontaneously displaced the other fluid. The non-wetting fluid spontaneously imbibed into the core, in case a wetting fluid contact angle of 133 - 180° was observed. It can therefore be concluded that spontaneous imbibition only occurs under the presence of a highly wetting imbibing fluid.

5.2 Spontaneous imbibition in carbonates

Due to the characteristic texture and pore network structure, spontaneous imbibition has a particular impact on the oil recovery of carbonates reservoirs. According to the dual-porosity model, highly permeable and conductive fractures are surrounding low permeable matrix systems. Although the carbonate matrix typically has a much lower conductivity, it contains the majority of the oil [105]. Since an aquifer drive or water injection follows the path of least resistance, the water typically bypasses the matrix. As a consequence, oil is predominantly displaced from the fracture systems. To recover oil from the matrix, it is therefore essential to obtain spontaneous matrix imbibition [74].

Depending on the fluid flow of the fracture-matrix system, it is distinguished between co-current and counter-current spontaneous imbibition. Both flow schemes are illustrated in Figure 5.2. While the oil and water phase are flowing into an identical direction during co-current imbibition, the two fluid phases enter/leave the rock matrix through different faces (Figure 5.2, a). In contrast, the counter-current imbibition describes the process in which the water and oil phases are flowing into opposite directions but enter/leave the matrix through the identical face (Figure 5.2, b) [74].

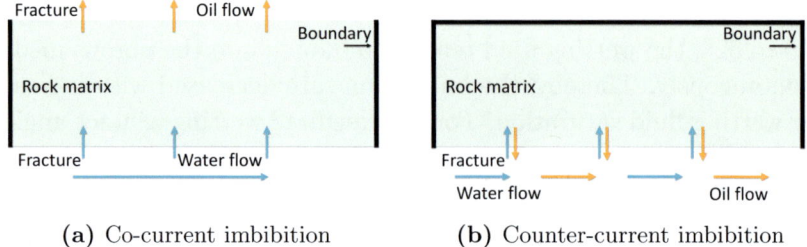

(a) Co-current imbibition (b) Counter-current imbibition

Figure 5.2: Co-current and counter-current imbibition scheme (redrawn after [87]).

Co-current and counter-current imbibition flow regimes are comprehensively studied on a wide range of geometries and boundary conditions. Thereby, counter-current imbibition is identified as the dominating flow regime during spontaneous imbibition. However, in regard to carbonate reservoirs, it is suggested that the characteristic fracture-matrix structure promotes co-current imbibition. In comparison to counter-current flow, co-current imbibition causes a faster and more efficient oil recovery [78].

5.3 Methodology

To gain an improved understanding of the mechanisms of spontaneous brine imbibition into limestones, the experimental study emphasizes the impact of both, connate as well as imbibing brine composition. The connate and imbibing water combinations of the Formation-water (salinity of $183.4\,g/l$), Sea-water ($43.8\,g/l$) and 100 times Diluted-sea-water ($0.4\,g/l$) are thereby divided into four test groups: Group I, Identical salinity of connate water and imbibing water, Group II and III, Imbibition of a lower salinity brine into a system at higher salinity and Group IV, Imbibition of a higher salinity brine into a system at lower salinity. For reproducibility, each selected brine con-

5.3 METHODOLOGY

figuration was tested on two core samples: Preferably a higher permeable and a less permeable limestone sample. All tested connate water and imbibing water combinations are summarized in Figure 5.3.

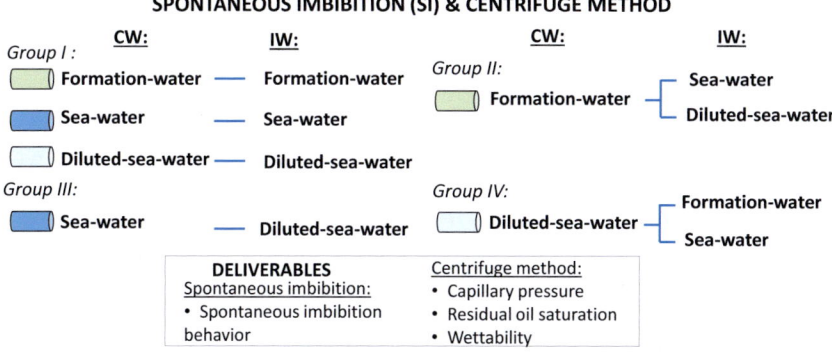

Figure 5.3: Spontaneous and forced imbibition methodology - The samples of the four test groups were initially exposed to spontaneous imbibition tests and then drained to So_r by using the centrifuge method.

Group I thereby evaluates the spontaneous imbibition behavior in case no composition difference between connate water (CW) and imbibing water (IW) is existing. It particularly examines the question, if the arising wettability can promote spontaneous imbibition under the absence of any composition difference/low-salinity-effect.

Group II and Group III are in line with the traditional low-salinity assumption, in which a high saline connate water and a lower saline imbibing water are combined. The two test groups are evaluating and comparing the potential of Sea-water and Diluted-sea-water to displace oil spontaneously.

Group IV reverses the connate water and imbibing water combinations of Group II/III. In this case, the connate water is represented by Diluted-sea-water, while Formation-water and Sea-water, respectively, are used as imbibing water.

5.4 Experimental conduction

After aging the cores and measuring the effective oil permeability at Sw_c, the samples were placed inside Amott cells and surrounded by the selected imbibing brine. Due to the temperature increase at the beginning of the experiment, the graduated part of the Amott cells was initially not entirely brine filled. Additionally, the top lid was only loosely tightened to avoid cell over-pressurization. No sign of evaporation or salt precipitation indications was observed. The set-up of the spontaneous imbibition experiments is illustrated in Figure 5.4.

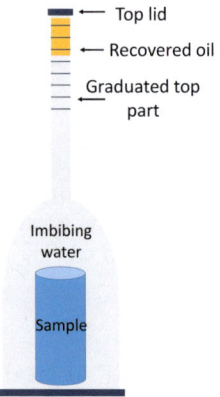

Figure 5.4: Schematic Amott imbibition cell set-up - The top lid allowed the pressure balancing during the heating up of the cells. The recovered oil accumulated at the graduated top part, which allowed a continuous production monitoring.

Once the imbibing water started to migrate into the cores, oil droplets accumulated at the top face of the core (cf. Figure 5.9). The arising oil bubbles remained attached to the core surface until the buoyancy forces caused the oil detachment and accumulation at the cell top. In order to accelerate the oil detachment from the core surface, the Amott cells were gently shaken every 24 $hours$. The spontaneous imbibition experiments were conducted at a temperature of 70°C over a period of approximately 20 $days$. The oil recovery was thereby

5.5 SPONTANEOUS IMBIBITION RESULTS

monitored as a function of time. Since the first oil recovery at the beginning of the experiment was generated by thermal expansion, the obtained oil production required a thermal expansion correction.

5.5 Spontaneous imbibition results

5.5.1 Identical salinity of connate water and imbibing water

Group I examines the spontaneous imbibition behavior in case the connate and imbibition brine are identical. Using Formation-water as connate water (CW) and imbibing water (IW), a marginal oil recovery of 1.9 % and 1.0 % was observed (cf. Table 5.2, Figure 5.5). The small oil production indicates oil wetting properties as the brine does not migrate into the cores. The spontaneous oil production of Sea-water as CW and IW resulted in 2.5 % and 2.1 % recovery, which is indicative of similar wetting conditions. Finally, Diluted-sea-water as CW and IW caused minor oil recovery of 3.1 % and 2.7 %. Figure 5.5 illustrates that independently of the brine composition and salinity, hardly any oil recovery was observed as long as the salinity of the connate and imbibing water was identical. All spontaneous imbibition results are summarized in Table 5.2.

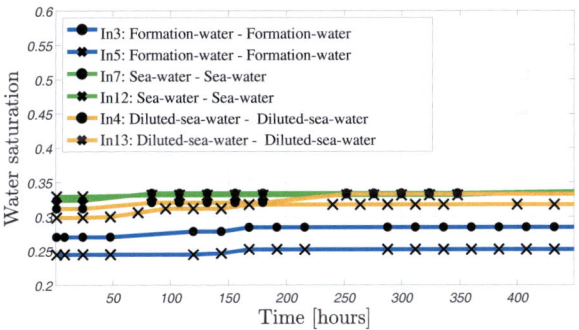

Figure 5.5: Spontaneous imbibition results Group I - Using the identical salinity of the connate water and imbibing water, hardly any spontaneous oil recovery was observed.

5.5 SPONTANEOUS IMBIBITION RESULTS

5.5.2 Imbibition of a low saline brine into a system at higher salinity (A)

Based on a common low-salinity definition, the samples of test Group II (Formation-water as CW) consist of a high-salinity connate water and lower salinity imbibing brine. An oil recovery of 25.4 % and 18.7 % was observed when using Formation-water as CW and Sea-water as IW (cf. Figure 5.6). Compared to Formation-water, the synthetic Sea-water is approximately 4 times less saline but has a significant sulfate concentration of $3.53\,g/l$. Therefore, the obtained production might be caused by the introduction of sulfate ions into the oil-brine-solid system and/or the reduction of the total salinity. The overall highest spontaneous oil recovery of 36.1 % and 34.7 % was observed for Formation-water as CW and the 400 times less saline Diluted-sea-water (sulfate concentration of $0.035\,g/l$) as IW. In this case, the spontaneous oil recovery is likely to be driven by the significant salinity difference between the connate and imbibing brine.

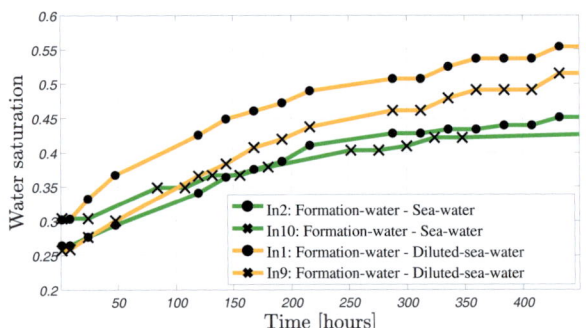

Figure 5.6: Spontaneous imbibition results Group II - Using Formation-water as CW, Diluted-sea-water as IW caused a 12.9% higher average oil recovery than Sea-water as IW.

5.5.3 Imbibition of a low saline brine into a system at higher salinity (B)

Similar to test Group II, Group III contains a high saline connate water (Sea-water) and a lower saline imbibing water (Diluted-sea-water). The spontaneous imbibition resulted in an oil recovery of 22.5 % and 22.3 % (cf. Figure 5.7), which is on average 13 % less oil recovery compared to Formation-water as CW and Diluted-sea-water as IW. The results are indicating, that the establishment of a significant salinity difference between connate and imbibing water is leading to higher oil recovery rather than the introduction/presence of sulfate ions into/inside the system.

5.5.4 Imbibition of a high saline brine into a system at lower salinity

To complete the possible CW and IW combinations, Group IV reverses the conventional low-sanity tests. Two cores were initially saturated in Diluted-sea-water and then imbibed by Formation-water and Sea-water. None of the cores showed oil production, as illustrated in Figure 5.8.

5.6 Contact angle

According to its fundamental physics, spontaneous imbibition is closely linked to capillary pressure and hence to the arising contact angle of the wetting fluid. A qualitative impression of the resulting contact angles is displayed in Figure 5.9.

Compared to Formation-water and Sea-water as IW (Figure 5.9, a, and Figure 5.9, b), Diluted-sea-water as IW is causing the strongest water-wet contact angle (Figure 5.9, c, Figure 5.9, d).

In addition to the qualitative contact angle impression, the Pendant-drop measurement system was modified to measure the arising contact angles quantitatively. After completing the spontaneous imbibition tests, the core samples were therefore immersed into a transport brine filled container. A bent needle syringe was used to generate

5.6 CONTACT ANGLE

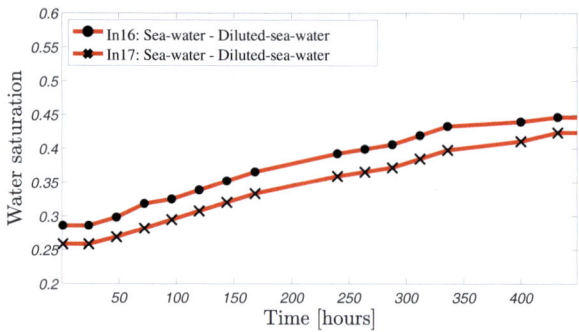

Figure 5.7: Spontaneous imbibition results Group III - The average spontaneous oil recovery of Sea-water (CW) and Diluted-sea-water (CW) is in the range of the spontaneous oil recovery of Formation-water (CW) and Sea-water (IW).

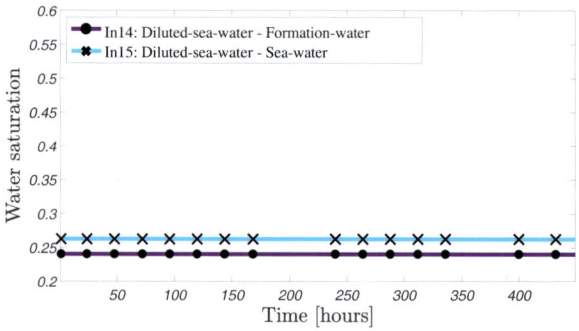

Figure 5.8: Spontaneous imbibition results Group IV - No oil recovery was observed in case of a lower saline connate water and a higher saline imbibing water.

a small oil bubble at the bottom face of the core, which was then captured by the optical camera system. The measurements of the Sea-water-oil-solid and Diluted-sea-water oil-solid-systems are displayed in Figure 5.10.

Although the differences between the measured contact angles are significant, the conducted contact angle measurement has some limitations. Initially, the core and wettability heterogeneity are complicating the acquisition of reproducible contact angle values. Further-

5.6 CONTACT ANGLE

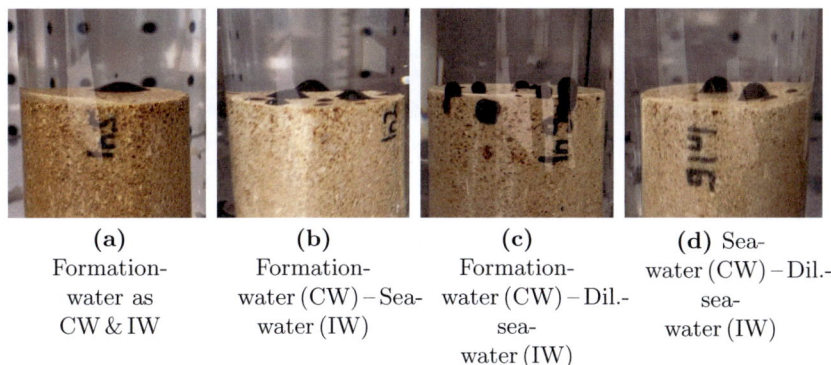

(a) Formation-water as CW & IW

(b) Formation-water (CW) – Sea-water (IW)

(c) Formation-water (CW) – Dil.-sea-water (IW)

(d) Sea-water (CW) – Dil.-sea-water (IW)

Figure 5.9: Spontaneous brine imbibition tests at 70°C - Compared to Formation-water and Sea-water as IW, the Diluted-sea-water system is characterized by the strongest water-wet contact angle.

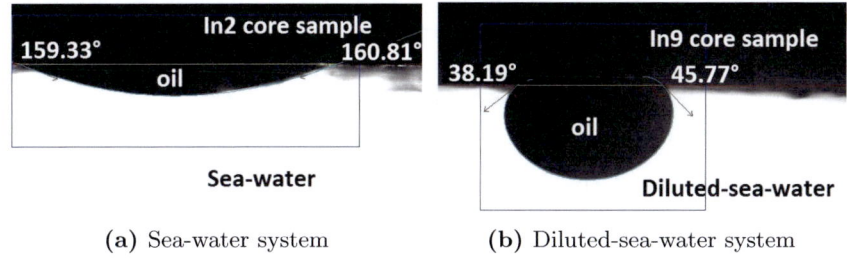

(a) Sea-water system

(b) Diluted-sea-water system

Figure 5.10: Contact angle measurements at ambient conditions - In accordance with the qualitative impression during the spontaneous imbibition tests, the Diluted-sea-water system resulted in the strongest water-wet contact angle.

more, the arsing flat oil droplet geometries of the Formation-water and Sea-water systems are hampering the optical measurements. However, besides the above mentioned drawbacks, the qualitative contact angle observations and quantitative contact angle measurement indicate that the presence of the Diluted-sea-water is leading into a stronger water-wet wettability than the presence of Sea-water or Formation-water.

5.7 Zeta potential

As summarized in the literature review (Chapter 3), Surface charge change is one of the proposed mechanisms of low-salinity effects in carbonates [64]. The mechanism assumes that the carbonate surface charge is impacted by several factors such a brine composition, salinity, pH value and temperature.

A comprehensive study on the salinity and pH value impact on zeta potential measurements in brine-carbonate systems was reported in the work of Mahani et al. [65]. At a neutral pH value of 7, the zeta potential of following limestone-brine suspensions were measured: Formation-water and limestone $\approx +5mV$, Sea-water and limestone $\approx -5\,mV$ and 25 times Diluted-sea-water and limestone $\approx -15\,mV$. Note, that the brine composition of the Formation-water, Sea-water and Diluted-sea-water of this thesis (*) and the work of Mahani et al. [65] are almost identical. The ionic strength of the brines were as following: Formation-water 3.655 (*) and 3.659 ([65]), Sea-water 0.870 (*) and 0.869 ([65]) and Diluted-sea-water 0.0087 (*) and 0.017 ([65]).

The decrease of the zeta potential/surface charge due to the introduction of a lower saline brine is a possible explanation of the obtained high spontaneous imbibition of Section 5.5. Introducing a lower saline brine into the system and hence changing the surface charge oil-brine-limestone towards stronger negative values, the concept of Surface charge change assumes the generation of stronger water-wetting properties. As a result of the arising water-wetness, acid oil components are stimulated to detach from the pore surface.

In order to evaluate Mahani et al. [65] results, the zeta potential of six brine-limestone suspensions were measured as a function of temperature. The sample preparation procedure was thereby in line with the proposed procedure of Mahani et al. [65]. Initially, a limestone sample was ball-milled, which allowed the subsequent mixture of $100\,ml$ of the selected brine and $1\,g$ of limestone powder. Before starting the zeta potential measurements, the suspensions were sonicated for $20\,minutes$ to break the intermolecular interactions.

5.7 ZETA POTENTIAL

After treating the suspensions inside the high-frequency sonciator and waiting for 24 $hours$, the zeta potential was measured within a Malvern zetasizer nano ZS.

While Mahani et al. [65] reported zeta potential values of Formation-water, Sea-water and Diluted-sea-water, this work only obtained stable and reproducible results until a maximum suspension conductivity of approximately 30 mS/cm. Above the conductivity threshold, the electrodes of the Malvern capillary cells almost immediately corroded. In order to elude the limitation of the used measurement set-up, the zeta potential of 3 times, 5 times, 10 times and 100 times Diluted-sea-water was measured. In line with the work of Mahani et al. [65] and the zeta potential reference values of +5 mV for Formation-water and limestone and -5 mV for Sea-water and limestone, Figure 5.11 confirms a correlation between decreasing the system's salinity and reducing zeta potential. At a temperature of 25°C, following zeta potential were measured: 3 times diluted Sea-water and limestone -5.5 mV, 5 times diluted Sea-water and limestone -6.5 mV, 10 times diluted Sea-water and limestone -8.7 mV and 100 times diluted Sea-water and limestone -16.6 mV (cf. Table 5.1). In addition to the Sea-water limestone system measurements, the zeta potential measurements of 20 times diluted Formation-water and limestone and 40 times diluted Formation-water and limestone resulted into -2.3 mV and -2.4,mV, respectively.

While the zeta potential measurements were limited to a maximum temperature of 45°C, Figure 5.11 indicates, that the zeta potential measurements are consistent within a temperature range of 25 to 45°C. It can hence be concluded that similar surface charges appear at the spontaneous imbibition test temperature of 70°C.

Although 20 times diluted Formation-water and 5 times diluted Sea-water are both characterized by a salinity of approximately 9 g/l, the measured zeta potential of 5 times diluted Sea-water and limestone is roughly 3 times more negative than the zeta potential of 20 times diluted Formation-water and limestone (cf. Table 5.1). The zeta potential difference might be caused by the concentration difference of the potential determining ions (PDI). The 20 times diluted

5.8 SUMMARY & CONCLUSIONS

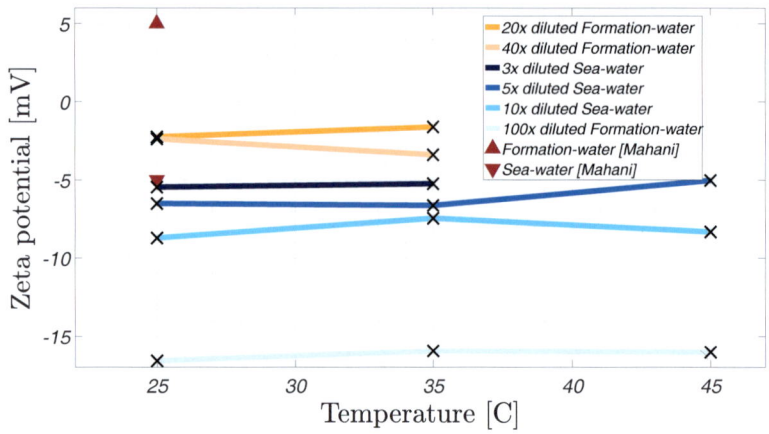

Figure 5.11: Zeta potential measurement results - The results indicate a correlation between the salinity of the brine and the system's zeta potential. Besides the experimentally acquired zeta potential values of the four tested brine-limestone systems, the Figure furthermore includes the high saline reference value of the work of Mahani et al. [65].

Formation-water has a Ca^{2+} concentration of $0.75\,g/l$, Mg^{2+} concentration of $0.1\,g/l$ and a SO_4^{2-} concentration of $0.02\,g/l$ while the 5 times diluted Sea-water is characterized by a Ca^{2+} concentration of $0.1\,g/l$, a Mg^{2+} concentration of $0.32\,g/l$ and a SO_4^{2-} ion concentration of $0.71\,g/l$. The results are in accordance with the work of Zhang et al. [111], which obtained positive zeta potential values when increasing the Mg^{2+} and Ca^{2+} ion concentration, and negative zeta potentials values when increasing the SO_4^{2-} ion concentration.

5.8 Summary & conclusions

Based on the comparison of the effective oil permeability measurements before and after aging, the core samples exhibited a wettability alteration towards intermediate to oil-wet wettability conditions.

In case the connate and imbibing water had the identical composition (Group I), a marginal spontaneous oil recovery of $1.5\,\%$, $2.3\,\%$

5.8 SUMMARY & CONCLUSIONS

	Salinity [g/l]	Zeta potential [mV]	Conductivity [mS/cm]	Zeta potential [mV]	Conductivity [mS/cm]	Zeta potential [mV]	Conductivity [mS/cm]
20 x dil. Form.-water	9.7	-2.3	20.2	-1.6	21.3	-	-
40 x dil. Form.-water	4.8	-2.4	9.7	-3.4	11.0	-	-
3 x dil. Sea-water	14.5	-5.5	24.6	-5.2	27.3	-	-
5 x dil. Sea-water	8.8	-6.5	15.9	-6.6	17.4	-5.0	18.9
10 x dil. Sea-water	4.4	-8.7	8.7	-7.5	9.7	-8.8	9.9
100 x dil. Sea-water	0.4	-16.6	1.1	-16.0	1.2	-16.0	1.4
		25°C		35°C		45°C	

Table 5.1: **Zeta potential measurement results** - The table summarizes the zeta potential and corresponding conductivity measurements of the six brine-limestone system at 25, 35 and 45°C.

and 2.9 % was observed for Formation-water, Sea-water and Diluted-sea-water, respectively. In the absence of a salinity/composition difference, the imbibing water is incapable of imbibing into the cores. According to the result of the idealized spontaneous imbibition tests on PTFE cores [79], the arising water contact angle is not small enough to cause the concave water oil-interface imbibition into the core.

The combination of a higher saline connate water and lower saline imbibing water (Group II and Group III) resulted in significant spontaneous oil recovery. Formation-water (CW) and Sea-water (IW) caused an average oil recovery of 22.1%, Formation-water (IW) and Diluted-sea-water (CW) caused an average recovery of 35.4% and Sea-water (CW) and Diluted-sea-water (IW) caused an average oil recovery of 22.4%. The results are indicating, that the establishment of a significant salinity difference between connate and imbibing water is leading to higher oil recovery rather than introduction/presence of sulfate ions into the system.

5.8 SUMMARY & CONCLUSIONS

Finally, the combination of a lower saline connate (IW) and a higher saline imbibing water did not cause any oil recovery (Group IV). It can therefore be concluded that spontaneous imbibition only occurs in case of a higher saline connate water and lower saline imbibing water. The results of the spontaneous imbibition tests are summarized in Table 5.2.

The literature typically links low-salinity effects to wettability alteration. Compared to Formation-water as IW and Sea-water as IW, the qualitative and quantitative contact angle observations showed the strongest water-wet behavior for Diluted-sea-water as IW.

The work of Zhang et al. [111] and Mahani et al. [64, 65] proposed Surface charge change as a possible mechanism of low-salinity effects in carbonates. Although the used zeta potential measurement set-up did not allow the zeta potential measurements of undiluted Formation-water and Sea-water systems, the step-wise brine dilution and the corresponding zeta values showed a clear tendency. The introduction of a lower saline brine is causing an alteration of the limestone surface charge towards stronger negative values. In line with the work of Mahani et al. [64, 65], stronger negative surface charges cause the acid oil component detachment from the pore surface. Consequently, the arising wettability alteration promotes spontaneous imbibition, which was observed when combining a high saline connate water and a low saline imbibing water.

The main findings of the spontaneous imbibition, contact angle and zeta potential experiments are as follows.

- Due to the geological characteristics of carbonates, spontaneous imbibition is an important oil recovery mechanism in carbonates.

- The spontaneous imbibition experiments resulted in oil recovery in case the imbibing water had a lower salinity than the connate water. Hardly any oil production was observed in case the imbibing water had a higher or the same salinity as the connate water.

5.8 SUMMARY & CONCLUSIONS

Group	Sample	CW	IW	Sw initial[%]	Sw spon. imbib. [%]	Oil recovery [%]	Average oil recovery [%]
I	In3	FB	FB	27.0	28.4	1.9	1.5
	In5	FB	FB	24.5	25.2	1.0	
	In7	SW	SW	32.4	34.0	2.5	2.3
	In12	SW	SW	32.9	34.3	2.1	
	In4	DSW	DSW	31.2	33.3	3.1	2.9
	In13	DSW	DSW	29.8	31.8	2.7	
II	In2	FB	SW	26.4	45.1	25.4	22.1
	In10	FB	SW	30.4	43.4	18.7	
	In1	FB	DSW	30.2	55.4	36.1	35.4
	In9	FB	DSW	25.7	51.5	34.7	
III	In16	SW	DSW	28.6	44.6	22.5	22.4
	In17	SW	DSW	25.8	42.4	22.3	
IV	In14	DSW	FB	24.0	24.0	0	0
	In15	DSW	SW	26.3	26.3	0	

Table 5.2: **Spontaneous imbibition results** - A detailed summary of the fourteen spontaneous imbibition experiments at 70°C. The highest spontaneous oil recovery was observed for Formation-water as CW and Diluted-sea-water as IW, followed by Sea-water (CW) and Diluted-sea-water (IW) and Formation-water (CW) and Sea-water (IW).

- Formation-water as CW and Sea-water as IW led into an average spontaneous oil recovery of 22.1 % while Formation-water and Diluted-sea-water resulted in an average spontaneous oil recovery of 35.4 %. The approximately 13.4 % higher spontaneous oil recovery of Diluted-sea-water as IW indicates the

5.8 SUMMARY & CONCLUSIONS

promising potential of Diluted-sea-water to efficiently recovery oil.

- The contact angle measurements indicate a wettability alteration towards stronger water-wet conditions when introducing Sea-water and/or Diluted-sea-water into the oil-brine-solid system.

- The zeta potential measurements of the different brine limestone system confirmed a correlation between the system's salinity and the limestone surface charge. As the salinity of the test brines decreases, the resulting zeta potential became more negative. This is in line with the spontaneous imbibition tests, which resulted in oil recovery in case a high saline connate water and low saline imbibing water were combined.

6
Centrifuge method

In a porous medium, capillary pressure defines the differential pressure between two immiscible fluid phases, which has to be overcome to initiate flow. Besides viscous and gravitational forces, capillary pressure significantly impacts the oil recovery of a reservoir [19]. In regard to carbonate reservoirs, capillary pressure has particular importance as it potentially promotes spontaneous oil recovery from the matrix system [74]. Centrifuge method, porous plate and mercury injection are common capillary pressure measurement methods [6, 72]. In this study, two ultra-centrifuges were used to acquire imbibition capillary pressure curves.

6.1 Theory

The characteristics of the centrifuge methods require an accurate and careful data interpretation. In line with the standard literature, this section develops the principle of the centrifuge method at the example of an (oil-brine) drainage experiment (Figure 6.1). Thereby, the brine saturated core samples are mounted inside a steel bucket, surrounded by crude oil and then spin around an axis of rotation. As a result of the arising centrifugal acceleration g_c, the (denser) brine inside the core is mobilized and flows towards the core outlet face (r_{outlet}). Oil enters at the core inlet face (r_{inlet}) and replaces

6.1 THEORY

the mobilized water. The corresponding (capillary) inlet pressure can be calculated depending on the centrifuge dimension, core dimension and fluid properties.

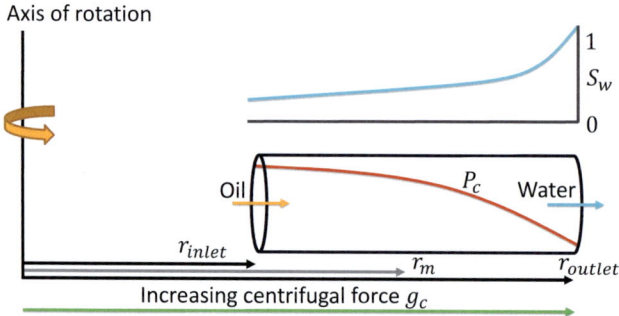

Figure 6.1: Primary drainage centrifuge schematics - The centrifugal acceleration causes the water movement into the opposite direction of the axis of rotation. The imbibing oil replaces the mobilized water.

While the imbibing oil efficiently replaces water around the inlet face, the water saturation remains higher towards the core outlet. Hassler & Brunner first formulated, that the outlet saturation of the wetting fluid remains at saturation of 100 % [49]. Since capillary pressure only occurs under the presence of two immiscible fluids, the outlet face capillary pressure consequently remains at zero.

Depending on the centrifuge design, the amount of replaced water at each centrifuge speed is either monitored by an optical or a manual measurement system. It is evident that the captured effluent quantity represents the average core fluid saturation and does not reflect the actual saturation heterogeneity. The centrifuge method therefore requires an analytical and/or numerical approach to correct the average saturation to the inlet saturation.

Since most analytical and numerical approaches are based on the Hassler & Brunner conditions (outlet saturation of 100% and outlet capillary pressure of zero), the compliance of the suggested conditions is crucial for the experimental data processing. Although several studies questioned the validity of the outlet boundary conditions, a breakthrough of the non-wetting fluid does not occur [38, 73].

In order to avoid the core samples desaturation, the design of a centrifuge experiment should furthermore respect the concept of critical Bond number. Generally, the Bond number N_B relates the occurring centrifugal forces to the capillary forces [72]

$$N_B = \frac{k \cdot \Delta\rho \cdot \omega^2 \cdot r_m}{\sigma}, \qquad (6.1)$$

where k is the permeability in $[m^2]$, $\Delta\rho$ is the density difference in $[kg/m^3]$, ω is the angular velocity in $[rad/s]$, r_m is the distance from the center of rotation to the middle of the core in $[m]$ and σ is the interfacial tension between the fluid phases in $[J/m^2]$.

As a rule of thumb, a Bond number/capillary number smaller than 10^{-5} describes a flow environment in which capillary pressure dominates the flow conditions. Therefore, once the residual oil has been reached, capillary pressure prevents the mobilization of further oil. The application of excessive gravitational force might overcome the capillary forces and hence cause core desaturation. In this case, the experimentally obtained residual oil saturation underestimates the actual residual oil saturation and is not representative of a reservoir scale oil recovery.

In regard to the non-water-wet carbonate samples, the comprehensive experimental study of Humphry et al. [53] suggested a critical Bond number of 10^{-4} to 10^{-3}. The maximum experimentally applied Bond numbers are summarized in Table 7.1. Except for sample In14, the corresponding maximum Bond number did not exceed a magnitude of 10^{-6}. Sample desaturation did not occur.

6.2 Methodology

The experimental spontaneous imbibition and centrifuge methodology is summarized in Figure 5.3. After completing the spontaneous

6.3 ANALYTICAL CENTRIFUGE SOLUTION

imbibition tests, the core samples were placed inside the centrifuge and consecutively drained to residual oil saturation. Thereby, the study investigates the connate water and imbibing water composition impact on residual oil saturation, capillary pressure and wettability.

The centrifuge method is a gravity-stable method and does not interfere with viscous fingering. Furthermore, the technique is commonly considered to be less impacted by capillary end effects [7, 72]. It is the recommended method to determine true residual oil saturation. Moreover, the centrifuge method directly measures and reflects capillary pressure.

The implemented analytical approach combines a hyperbolic fit of the experimentally obtained data [30] and Forbes first solution [37] to determine the inlet water saturation. A comprehensive description of the method is provided in Section 6.3. A numerical centrifuge model is developed in Chapter 8 to validate the experimentally obtained data.

6.3 Analytical centrifuge solution

The heterogeneous water saturation profile of Figure 6.1 illustrates a drawback of the centrifuge method. It is obvious that the experimentally captured effluent production does not reflect the actual saturation heterogeneity. Furthermore, the arising capillary pressure can only be calculated at the core inlet face. Therefore, the data analysis of the centrifuge method requires an analytical and/or numerical approach to relate the experimentally obtained average water saturation to the inlet capillary pressure/water saturation.

The first analytical average to inlet saturation correction was proposed by Hassler & Brunner in 1945 [49]. The obtained average water saturation along the core can be summarized as following [76]

$$Sw_{av} = \frac{1}{(r_{outlet} - r_{inlet})} \int_{r_{inlet}}^{r_{outlet}} Sw(r)dr, \qquad (6.2)$$

6.3 ANALYTICAL CENTRIFUGE SOLUTION

where Sw_{av} is the average water saturation, r_{inlet} is the distance of the axis of rotation to the core inlet, r_{outlet} is the distance of the axis of rotation to the core outlet and Sw is the local water saturation. Assuming the water saturation as a function of the locally occurring capillary pressure (P_c=$\rho g_c L$), the average water saturation at any point (r) along the core may be rewritten according to Equation 6.3 [101]

$$Sw_{av} = \frac{1}{\rho g_c (r_{outlet} - r_{inlet})} \int_{r_{inlet}}^{r_{outlet}} Sw(\rho g_c L) dr, \qquad (6.3)$$

where ρ is the fluid density, g_c is the centrifugal acceleration and L denotes the core length. Equation 6.3 emphasizes the difficulty of correcting the average water saturation to the local water saturation. Since the water saturation, capillary pressure and centrifugal force are a function of the distance to the axis of rotation and interfering with each other, the exact mathematical solution of the average saturation (Equation 6.3) yields into a Volterra equation [76]. To avoid a Volterra equation, different approaches have been proposed. Hassler & Brunner first suggested that in case of a drainage process, the outlet wetting fluid saturation remains at a saturation of 100 %. Consequently, the outlet capillary pressure is zero.

In this case, the total pressure along the core can be simplified to Equation 6.4 [49, 101]

$$P_c = \rho g_c L. \qquad (6.4)$$

Furthermore, Hassler & Brunner demonstrated that r_{inlet}=r_{outlet} can be assumed in case the core length becomes negligible small in relation to the length of the rotation axis. This is valid for r_{inlet}/r_{outlet} ratios which are larger than 0.7. Equation 6.3 can then be rewritten as a function of the capillary inlet pressure (P_{ci})

$$Sw_{av} = \frac{1}{P_{ci}} \int Sw_i dP_{ci}, \qquad (6.5)$$

6.3 ANALYTICAL CENTRIFUGE SOLUTION

where Sw_i is the inlet water saturation. Differentiation and rearranging leads to Equation 6.6

$$Sw_i = Sw_{av} + P_{ci}\frac{dSw_{av}}{P_{ci}}. \tag{6.6}$$

Equation 6.6 is known as the Hassler & Brunner average to inlet water correction [49, 101]. Forbes [37] proposed to add a geometrical correction factor, which is widely acknowledged to improve the Hassler & Brunner solution [72]

$$Sw_i = Sw_{av} + \frac{1}{1+\alpha} P_{ci} \frac{dSw_{av}}{dP_{ci}}, \tag{6.7}$$

where the parameter α depends on the centrifuge and core dimensions

$$\alpha = \frac{1 - \sqrt{1-b}}{1 + 2\sqrt{1-b}}, \quad \beta = 1 - \left(\frac{r_{inlet}}{r_{outlet}}\right)^2. \tag{6.8}$$

The corresponding capillary inlet pressure can be directly calculated based on the fluid density, centrifuge and core dimensions

$$\begin{aligned} P_{ci} &= \frac{1}{2} \cdot \Delta\rho \cdot \omega^2 \cdot \left(r_{outlet}^2 - r_{inlet}^2\right) \\ &= \frac{1}{2} \cdot \Delta\rho \cdot \left(\frac{RPM \cdot 2\pi}{60}\right) \cdot \left(r_{outlet}^2 - r_{inlet}^2\right), \end{aligned} \tag{6.9}$$

where $\Delta\rho$ is the density difference between the oil and water phase in $[kg/m^3]$, ω is the angular velocity of the centrifuge in $[rad/s]$ and RPM denotes the centrifuge spin velocity in revolutions per minute.

Forbes first solution (Equation 6.7) can be directly solved by the derivative of the average saturation with respect to the capillary pressure by determining the slope between the experimentally

6.3 ANALYTICAL CENTRIFUGE SOLUTION

determined data. This study, however, uses an alternative approach by initially conducting a hyperbolic regression of the experimentally obtained data. The applied method has two main advantages. Besides smoothing the experimental data, the arising mathematical function is used to determine an exact solution of the average water saturation/capillary pressure derivative. The method was initially suggested by Donaldson et al. [30] by applying the least square solution of a three constant hyperbolic function to fit capillary drainage experiments

$$P_{ci} = \frac{B + C \cdot Sw_{av}}{1 + D \cdot Sw_{av}}, \qquad (6.10)$$

where the constants B, C and D are the constants of the least square solution of the experimentally obtained average water saturation and capillary pressure data. At first glance, the application of the method might not be straightforward. The application of the approach is therefore demonstrated at the example of core sample In17.

$$num(1) = \sum Sw_{av}^2 \cdot \left(\sum Sw_{av} P_c \cdot \sum Sw_{av} P_c^2 - \sum P_c \cdot \sum Sw_{av}^2 P_c^2 \right)$$
$$+ \sum Sw_{av} P_c \cdot \left(\sum Sw_{av} \cdot \sum Sw_{av}^2 P_c^2 - \sum Sw_{av} P_c \cdot \sum Sw_{av}^2 P_c \right) \quad (6.11)$$
$$+ \sum Sw_{av}^2 P_c \left(\sum P_c \cdot \sum Sw_{av}^2 P_c - \sum Sw_{av} \cdot \sum Sw_{av} P_c^2 \right)$$

$$num(2) = n \cdot \left(\sum Sw_{av}^2 P_c \cdot \sum Sw_{av} P_c^2 - \sum Sw_{av} P_c \cdot \sum Sw_{av}^2 P_c^2 \right)$$
$$+ \sum Sw_{av} \cdot \left(\sum P_c \cdot \sum Sw_{av}^2 P_c^2 - \sum Sw_{av} P_c \cdot \sum Sw_{av} P_c^2 \right) \quad (6.12)$$
$$+ \sum Sw_{av} P_c \left(\sum Sw_{av} P_c \cdot \sum Sw_{av} P_c - \sum P_c \cdot \sum Sw_{av}^2 P_c \right)$$

6.3 ANALYTICAL CENTRIFUGE SOLUTION

$$num(3) = n \cdot \left(\sum Sw_{av}^2 \cdot \sum Sw_{av} P_c^2 - \sum Sw_{av} P_c \cdot \sum Sw_{av}^2 P_c \right)$$
$$+ \sum Sw_{av} \cdot \left(\sum P_c \cdot \sum Sw_{av}^2 P_c - \sum Sw_{av} \cdot \sum Sw_{av} P_c^2 \right) \quad (6.13)$$
$$+ \sum Sw_{av} P_c \left(\sum Sw_{av} \cdot \sum Sw_{av} P_c - \sum P_c \cdot \sum Sw_{av}^2 \right)$$

$$denom = n \cdot \left(\sum Sw_{av}^2 P_c \cdot \sum Sw_{av}^2 P_c - \sum Sw_{av}^2 \cdot \sum Sw_{av}^2 P_c^2 \right)$$
$$+ \sum Sw_{av} \cdot \left(\sum Sw_{av} \cdot \sum Sw_{av}^2 P_c^2 - \sum Sw_{av} P_c \cdot \sum Sw_{av}^2 P_c \right) \quad (6.14)$$
$$+ \sum Sw_{av} P_c \left(\sum Sw_{av} P_c \cdot \sum Sw_{av}^2 - \sum Sw_{av} \cdot \sum Sw_{av}^2 P_c \right)$$

$$B = \frac{num(1)}{denom} \quad C = \frac{num(2)}{denom} \quad D = \frac{num(3)}{denom} \quad (6.15)$$

Table 6.1 summarizes the step-wise calculation of the combined hyperbolic regression and Forbes first solution inlet water saturation calculation. Initially, the required quantities of Equation 6.11 to 6.14 are calculated to prepare the calculation of Equation 6.15. Once the least square solution constants B, C and D of the hyperbolic function are determined, the linear regressed capillary inlet pressure P_{ci} is calculated based on Equation 6.10. A comparison of the raw and regressed capillary pressure inlet data is plotted in Figure 6.2. It is shown that the obtained hyperbolic function precisely regresses the experimental raw data.

Since the experimentally obtained average water saturation and the corresponding capillary inlet pressure is described by a mathematical function, Forbes first solution (Equation 6.7) can be solved by inserting the exact solution of the average water saturation/capillary pressure derivative. Therefore Equation 6.10 is differentiated and substituted into Equation 6.7. The resulting Equation 6.16 combines the regressed experimental data with Forbes first solution to obtain a precise and unified water inlet solution calculation

6.3 ANALYTICAL CENTRIFUGE SOLUTION

Sw_{av}	P_c	$Sw_{av}P_c$	Sw_{av}^2	P_c^2	$Sw_{av}P_c^2$	$Sw_{av}^2 P_c$	$Sw_{av}^2 P_c^2$
0.57	-0.13	-0.07	0.33	0.02	0.01	-0.04	0.01
0.61	-0.21	-0.13	0.37	0.04	0.03	-0.08	0.02
0.64	-0.35	-0.23	0.41	0.12	0.08	-0.14	0.05
0.68	-0.57	-0.39	0.47	0.33	0.22	-0.27	0.15
0.73	-0.94	-0.69	0.54	0.88	0.65	-0.51	0.47
0.78	-1.55	-1.20	0.60	2.40	1.87	-0.93	1.45
0.81	-2.55	-2.08	0.66	6.50	5.29	-1.69	4.31
0.85	-4.18	-3.54	0.72	17.49	14.79	-2.99	12.50
0.87	-7.24	-6.32	0.76	52.42	45.78	-5.52	39.98
$\sum Sw_{av}$	$\sum P_c$	$\sum Sw_{av}P_c$	$\sum Sw_{av}^2$	$\sum P_c^2$	$\sum Sw_{av}P_c^2$	$\sum Sw_{av}^2 P_c$	$\sum Sw_{av}^2 P_c^2$
6.55	-17.72	-14.65	4.86	80.20	68.71	-12.17	58.94

RPM [-]	P_{ci}	Sw_i	N	9
1000	-0.10	0.604	NUM(1)	-4.42
1280	-0.22	0.661	NUM(2)	8.35
1650	-0.36	0.714	NUM(3)	9.63
2110	-0.59	0.767	Denom	-8.80
2700	-0.99	0.820	B	0.50
3470	-1.57	0.856	C	-0.95
4450	-2.48	0.880	D	-1.09
5700	-4.03	0.896	α	0.48
7500	-7.40	0.906	β	0.11

Table 6.1: Hyperbolic regression of centrifuge imbibition data - The forced imbibition data of sample In17 is used to illustrate the calculation of the hyperbolic B, C and D parameters.

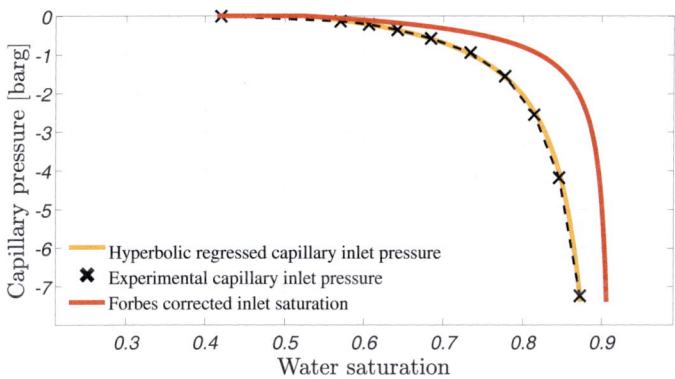

Figure 6.2: Comparison of average and inlet water saturation - The hyperbolic function precisely regresses the experimental data. The discrepancy between the average and inlet water saturation decreases with increasing centrifuge spin.

6.4 EXPERIMENTAL CONDUCTION

$$Sw_i = Sw_{av} + \frac{1}{1+\alpha} \cdot P_{ci} \cdot \frac{(1 + DSw_{av})^2}{(C - BD)}. \qquad (6.16)$$

The experimental average water saturation and derived inlet water saturation is plotted in Figure 6.2. Especially at lower centrifuge spins, the discrepancy between the saturation values is significant. As the centrifuge spin increases, the average water saturation approaches the inlet water saturation. While Donaldson et al. [31] focused their work on the analytical analysis of drainage curves, this study confirms the applicability of the method on imbibition experiments. Furthermore, the results are demonstrating that Donaldson's method can be combined with Forbes first solution.

6.4 Experimental conduction

Since its invention in 1945, especially the development of automatized saturation monitoring systems significantly improved the applicability and reliability of the centrifuge method. While older centrifuge systems required the centrifuge run interruption to determine the effluent production, state-of-the-art systems provide an in-situ saturation monitoring. The two used Core Laboratories Beckmann ultra-centrifuges allow the simultaneous mounting and running of three $5\,cm$ long and $3.7\,cm$ thick core samples. The centrifuge can be operated at a spin velocity between 1000 to $16000\,RPM$ and a maximum temperature of $90\,°C$. In line with standard literature, the fundamentals of the centrifuge method were introduced based on the primary drainage process. Note that from this point onward, the centrifuge method refers to imbibition processes.

The imbibition centrifuge experiments were conducted after completing the spontaneous imbibition. After placing the samples inside the steel bucket and mounting the transparent receiving tube, the imbibing brine was immediately added to avoid salt precipitation (cf. Figure 6.3). A known amount of oil V_o^{init} was then added to ensure a sharp and immediate oil-water interface from the beginning of the centrifuge run. The centrifuge buckets and rotor were preheated in-

6.4 EXPERIMENTAL CONDUCTION

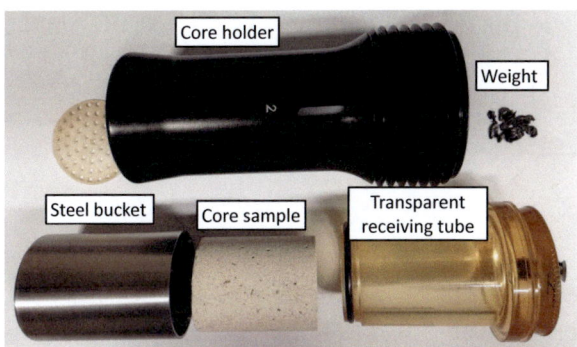

Figure 6.3: Imbibition centrifuge core holder set-up - The set-up consists of a steel bucket, a transparent receiving tube and a core holder. The metal weight balances possible weight differences.

side an external oven, mounted and assembled into the centrifuge. Each centrifuge run required the assembling of three sample holders to avoid the imbalance of the centrifuge rotor. Possible weight differences were furthermore corrected by placing metal weights on the top of the holders.

Overall nine rotation speeds in the range of 1000 to 7500 RPM were specified to ensure a dense capillary pressure measurement. Compared to the primary drainage, much lower rotations speeds are required as the imbibition centrifuge arm is longer. Each imbibition step was run for at least 30 $hours$.

The in-situ average water saturation was measured by a high definition camera. Thereby, the transparent receiving tube was illuminated by a strobe light which ensured the continuous fluid level imaging. The centrifuge software automatically converted the picture into pixel values. After starting the centrifuge, the camera captured the initial oil-brine interface position I_{init}. As a result of oil production, the oil-water interface I_{insitu} advanced towards the center of rotation. Consequently, the difference between the initial oil-water interface I_{init} and the final pixel interface I_{final} represented the total amount of produced oil in pixel equivalents. After completing the centrifuge run, the volume of the produced oil was determined

6.5 CENTRIFUGE RESULTS

with respect to the experimentally captured oil V_o^{cap} and the initially added oil V_o^{init}. Based on the amount of produced oil and pixels, a pixel factor PF was calculated

$$PF = \frac{V_o^{cap} - V_o^{init}}{I_{init} - I_{final}}. \tag{6.17}$$

Besides calculating the in-situ average water saturation, the obtained pixel factor indicates the reliability of the centrifuge experiment. During the study, each pixel factor was found to be close to 0.003. Significant deviations either result in an over- or underestimation of the saturation. Based on the pixel factor, the average water saturation at any point was determined by Equation 6.18

$$Sw_{av} = \frac{PF \cdot (I_{init} - I_{insitu}) + V_P \cdot Sw_c}{V_P}, \tag{6.18}$$

where Sw_{av} is the average water saturation and V_P is the pore volume. The obtained average water saturations were converted into the inlet water saturation by an analytical (Section 6.3) and numerical approach (Chapter 8).

6.5 Centrifuge results

For the first time, the implemented study reported centrifuge experiments to investigate the impact of brine composition and concentration on imbibition capillary pressure curve and residual oil saturation in limestones. The obtained average residual oil saturations ranged between 7.5 to 19.8 %, which is characteristic for mixed to oil-wet carbonates as reported, for example, by Masalmeh et al. [68].

The experimentally acquired data were initially smoothed by applying the least square solution of a three constant hyperbolic as proposed by Donaldson et al. [30] and subsequently corrected by Forbes first solution [39] (Equation 6.16). The analytically corrected imbibition capillary pressure curves of the experiments are displayed

6.5 CENTRIFUGE RESULTS

in Figure 6.4 to Figure 6.7, where the capillary imbibition curves are plotted on two different scales. Figure 6.4, a to Figure 6.7, a initially focus on a capillary pressure range of 0 to -3 $barg$. While this pressure range is more representative for reservoir scale related displacement processes, Figure 6.4, b to Figure 6.7, b denote the complete capillary imbibition experiments.

6.5.1 Identical salinity of connate water and imbibing water

During the spontaneous imbibition tests, all experiments which were performed using the same CW and IW showed non-water-wet behavior as the oil recovery ranged between 1 to 3.1 %.

In contrast to the spontaneous imbibition tests, the forced imbibition results of Group I (identical CW and IW) demonstrate a correlation between the system's salinity and wetting behavior. The imbibition capillary pressure curve of the six samples showed a decreasing oil-wetting tendency (less negative capillary pressure), as the salinity of the CW and IW combinations decreased. In accordance, the endpoint of the capillary pressure curves emphasized the effect of brine composition on the residual oil saturation. Formation-water as CW and IW resulted in a residual oil saturation (So_r) of 16.6 % and 13.9 %, Sea-water as CW and IW resulted into a So_r of 7.9 % and 6.9 % and Diluted-sea-water as CW and IW resulted into a So_r of 2.8 % and 3.6 %, respectively. Note that all residual saturation values are related to the Forbes corrected inlet water saturation.

In regard to the reported residual oil saturation values, the applied Forbes correction likely caused an underestimation of the residual oil (inlet) saturations. Especially for Diluted-sea-water as CW and IW, the discrepancy between average and inlet saturations is significant (4.6% and 4.1%).

6.5 CENTRIFUGE RESULTS

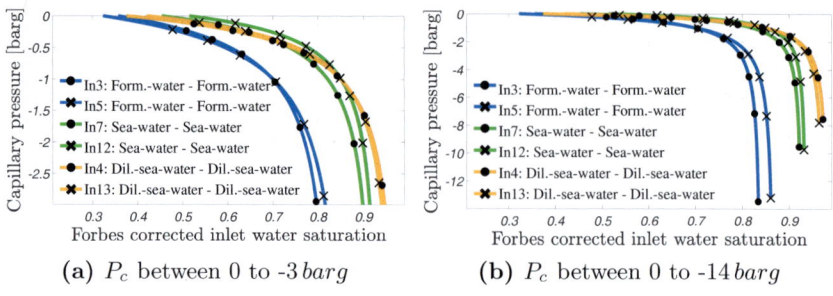

Figure 6.4: Forced imbibition results Group I - The imbibition capillary pressure curves are showing a decreasing oil-wetting tendency (less negative capillary pressure), as the salinity of the CW and IW combinations decreases.

6.5.2 Injection of a low saline brine into a system at higher salinity (A)

Due to the previous high spontaneous brine imbibition, the forced imbibition of Formation-water as CW and Sea-water and Diluted-sea-water as IW started at much higher water saturations around 43.4 to 55.4 %. Compared to the imbibition of the high saline Formation-water, both lower saline combinations showed less oil-wetting imbibition capillary pressure curves (i.e., less negative). Furthermore, the residual oil saturation of Formation-water as CW and Sea-water as IW were 8.7 % and 12.6 %, respectively and Formation-water (CW) and Diluted-sea-water (IW) were 4.8 % and 6.3 %, respectively (cf. Figure 6.5 and Table 6.2).

6.5.3 Injection of a low saline brine into a system at higher salinity (B)

The results of the forced imbibition of Sea-water (CW) and Diluted-sea-water (IW) are displayed in Figure 6.6. Since the obtained residual oil saturations of 11.8 % and 9.4 % were in the range of Formation-water (CW) and Sea-water (IW), the results indicate, that the establishment of a large salinity difference between CW and IW maximizes

6.5 CENTRIFUGE RESULTS

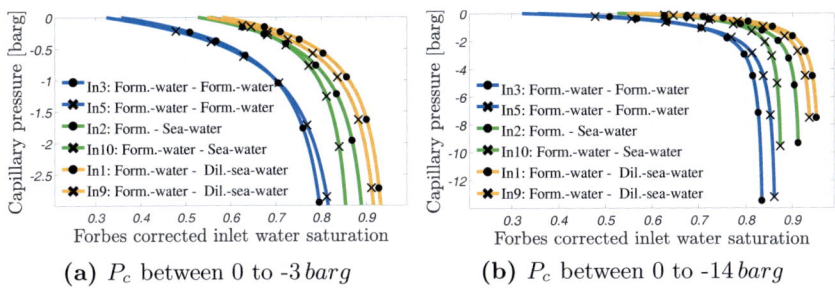

(a) P_c between 0 to $-3\,barg$ (b) P_c between 0 to $-14\,barg$

Figure 6.5: Forced imbibition results Group II plus In3 and In5 - In the case of Formation-water as CW, Diluted-sea-water as IW is leading into the strongest water-wetting behavior and smallest S_{o_r}.

the oil recovery. A similar tendency was obtained in the spontaneous imbibition section. The experimental data of In17 were used to validate the presented analytical and numerical centrifuge approaches (cf. Chapter 8).

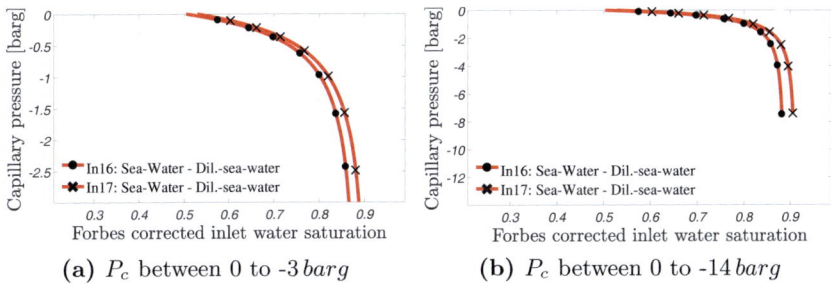

(a) P_c between 0 to $-3\,barg$ (b) P_c between 0 to $-14\,barg$

Figure 6.6: Forced imbibition results Group III - The forced imbibition results of Sea-water (CW) and Diluted-sea-water (IW) are in the range of Formation-water (CW) and Sea-water (IW).

6.5.4 Injection of a high saline brine into a system at lower salinity

The combination of a low saline connate water and higher saline imbibition brine did not cause any oil recovery during the sponta-

6.6 SUMMARY & CONCLUSIONS

neous imbibition. The residual oil saturation measured in the case of Diluted-sea-water (CW) and Formation-water (IW) was 16.6 % and for Diluted-sea-water (CW) and Sea-water (IW) was 10.6 % (cf. Figure 6.7).

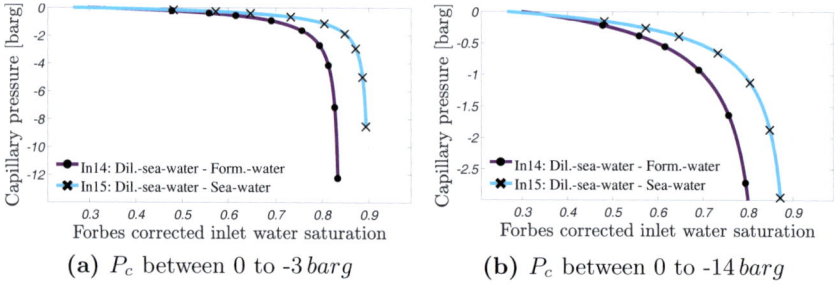

(a) P_c between 0 to -3 $barg$

(b) P_c between 0 to -14 $barg$

Figure 6.7: Forced imbibition results Group IV - Formation-water as imbibing water resulted into highest residual oil saturation.

6.6 Summary & conclusions

The reported capillary pressure curves showed a correlation between the system's salinity and the resulting wettability and residual oil saturation.

Under the absence of a brine composition difference, hardly any spontaneous oil recovery was observed. Although Diluted-sea-water as CW and IW did not show any spontaneous oil recovery, the combination resulted in the lowest residual oil saturation and strongest water-wet wettability. It underlines the assumption of Masalmeh et al. [69], that spontaneous imbibition tests do not necessarily reveal evidence about Enhanced oil recovery potential.

In the case of Formation-water as CW, Diluted-sea-water as IW caused the highest spontaneous imbibition as well as lowest residual oil saturation. Therefore, the spontaneous imbibition, zeta potential and centrifuge method experiments confirmed the promising potential of highly Diluted-sea-water to recover oil efficiently. Compared to the introduction of potential determining ion enriched brines, re-

6.6 SUMMARY & CONCLUSIONS

Group	Sample	Connate water	Imbibing water	S_w initial [%]	S_w spon. imbibition [%]	$S_{w,av}$ forced imbibition [%]	Forbes corrected $S_{w,inlet}$ [%]	Forbes corrected S_{or} [%]	Avg. Forbes S_{or} [%]	Oil recovery [%]	Maximum Bond number
I	In3	FB	FB	27.0	28.4	80.5	83.4	16.6	15.3	76.8	7.3E-6
	In5	FB	FB	24.5	25.2	82.6	86.1	13.9		81.4	9.2E-6
	In7	SW	SW	32.4	34.0	89.1	92.1	7.9	7.4	88.0	2.9E-6
	In12	SW	SW	32.9	34.3	90.6	93.1	6.9		89.5	5.5E-6
	In4	DSW	DSW	31.2	33.3	92.8	97.2	2.8	3.2	95.8	4.4E-6
	In13	DSW	DSW	29.8	31.8	92.3	96.4	3.6		94.7	9.4E-6
II	In2	FB	SW	26.4	45.1	88.3	91.3	8.7	10.7	84.1	9.2E-6
	In10	FB	SW	30.4	43.4	85.0	87.4	12.6		77.7	3.8E-6
	In1	FB	DSW	30.2	55.4	91.9	95.2	4.8	5.6	89.2	6.2E-6
	In9	FB	DSW	25.7	51.5	90.2	93.7	6.3		87.0	6.5E-6
III	In16	SW	DSW	28.6	44.6	85.2	88.2	11.8	10.6	78.7	2.7E-6
	In17	SW	DSW	25.8	42.4	87.4	90.6	9.4		83.7	8.0E-6
IV	In14	DSW	FB	24.0	24.0	80.2	83.4	16.6	16.6	78.1	1.3E-5
	In15	DSW	SW	26.3	26.3	85.7	89.4	10.6	10.6	85.6	3.9E-6

Table 6.2: **Centrifuge method results** - A detailed summary of the fourteen forced imbibition experiments at 70°C.

ducing the system's salinity might be a more promising low-salinity approach in limestones. In line with the conclusions Romanuka et al. [89], this assumption is furthermore emphasized by the obtained residual oil saturation of 4.8 % and 6.3 % for Formation-water as CW and Diluted-sea-water as IW.

Reversing the traditional low-salinity concept (Group IV) by combining a low saline connate water and a high saline imbibing water causes an inefficient oil recovery. The main conclusions of the experimental centrifuge part are the following.

- The experimental conduction of the centrifuge experiments was described in detail. The Hassler & Brunner correction was de-

6.6 SUMMARY & CONCLUSIONS

rived to illustrate the necessity of conducting a reliable inlet water saturation correction.

- The applied combination of the hyperbolic function and Forbes first solution is leading into an efficient, unified and precise analytical analysis of the imbibition data.

- The imbibition capillary pressure curves are characterized by an increasing water-wetting tendency and a simultaneous reduction of the residual oil saturation, as the salinity of the imbibition brines decreases in comparison to Formation-water.

- Moreover, the forced imbibition centrifuge experiments demonstrated the impact of brine composition on residual oil saturation. Using Formation-water as CW, Formation-water (IW), Sea-water (IW) and Diluted-sea-water (IW) caused a final Forbes corrected residual oil saturation of 15.3 %, 10.7 % and 5.6 %, respectively. The highest oil recovery was observed when Diluted-sea-water was used as CW and IW (So_r of 3.2 %).

7

Coreflooding

After testing the brine composition impact on spontaneous imbibition and capillary pressure behavior, the experimental study was completed by the conduction of three unsteady state experiments.

7.1 Theory

Compared to spontaneous imbibition, centrifuge method and steady state corefloodings, unsteady state (USS) corefloodings are the most representative laboratory reproduction technique of reservoir displacement processes [72]. Since the steady state method includes the simultaneous injection of two fluids, the method does not develop a proper displacement front. Furthermore, the centrifugal acceleration of the centrifuge method causes an oil-water replacement, which is not representative of a reservoir displacement process.

In its most general form, a USS coreflooding is described by a single-phase fluid injection at the inlet face. As a result of the displacing fluid injection, the displaced fluid is mobilized and is pushed towards the core outlet. Until the breakthrough of the displacing fluid, only the displaced fluid is produced at the core outlet [72].

Depending on the experimental objectives, the design and conduction of unsteady state corefloodings can significantly vary. Besides the displacing and displaced fluid properties, especially the ex-

7.1 THEORY

perimental conditions, quantity and duration of the selected injection rates depend on the experimental purpose.

This study includes the conduction of three USS corefloodings to test the secondary and tertiary displacement efficiency of Formation-water, Sea-water and Diluted-sea-water. To resemble field realistic flow conditions, each brine was initially injected at a field rate equivalent. Thereby, a representative field flow velocity is converted into a volumetric laboratory injection rate. While the study of Nasralla et al. [82] assumed a field flow velocity of $1\,feet/day$, the work of Kamath et al. [58] proposed a flow velocity of $0.5\,feet/day$. Assuming a field velocity of $1\,feet/day$ and depending on the core properties, the volumetric laboratory injection rate equivalent was calculated by using Equation 7.1

$$q_{lab} = v_f \cdot A \cdot \phi \cdot 0.3048 \cdot 10^{-6}, \tag{7.1}$$

where q_{lab} is the volumetric laboratory injection rate in $[cm^3/min]$, v_f is the field flow velocity in $[feet/day]$, A is the average core-cross section in $[m^2]$ and ϕ is the porosity.

While the injection at field rate equivalent results into a reservoir scale related flow environment, the practical implementation is time and labor-consuming. Once the displacing fluid is injected into the core, the displacement process causes an immediate and significant increase of differential pressure. Typically, the injection of several pore volume equivalents is required to obtain a pressure re-equilibrium. Depending on the sample properties, stabilized flow conditions are obtained after several hours or days of fluid injection.

In the case of very low injection rates, the corresponding differential pressure of highly permeable cores furthermore remains below the sensitivity of many pressure transducers. Consequently, the inaccuracy of the acquired pressure readings complicates the experimental data interpretation.

Besides the practical challenges of field rate equivalent injection applications, low injection rates are suspected of promoting laboratory artifacts. Commonly referred to as the capillary end effect,

7.1 THEORY

this phenomena describes the accumulation of the wetting fluid at the core outlet. During the simultaneous flow of two or more fluids, end effects are caused by the sharp transition between the capillary pressure and non-capillary pressure dominated flow environments. As a result of the capillary pressure discontinuity, the wetting fluid is trapped in the vicinity of the porous media outlet face. While capillary end effects cause an unrepresentative high wetting fluid accumulation, the non-wetting fluid approaches its residual saturation [46].

Figure 7.1 illustrates the impact of the capillary end effect on unsteady state corefloodings under the assumption of an oil-wet system. In regard to the short core length and expected oil-wet wettability of the limestones samples, the three conducted corefloodings are likely affected by the end effects. Consequently, the obtained oil recovery at field rate injection equivalent overestimates the actual remaining oil saturation [68].

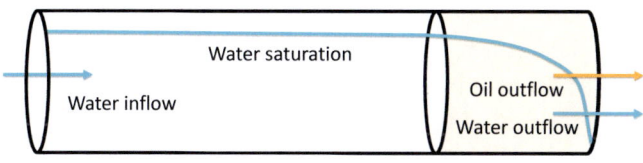

Figure 7.1: Capillary end effect - In the case of an oil-wet system, the capillary end effect prevents the oil phase from exiting the core (redrawn after [44] and [77]).

In tertiary injection mode experiments, capillary end effects furthermore distort the interpretation of additional oil recovery. Once the corefloodings change from secondary to tertiary injection, many experimental studies report additional oil recovery. However, the obtained oil recovery might be predominantly caused by the mitigation of end effects. In this case, the additional oil recovery is not representative of field oil recovery [97].

To mitigate the capillary end effect impact on experimental results, the injection rates are typically bumped up. Thereby, the

7.1 THEORY

arising faster flow velocities cause a stronger viscous dominated flow and hence a reduction of end effects [68].

It is, however, evident that the injection rates cannot infinitely be bumped up. In line with the centrifugal force of the centrifuge method, high injection rates can lead to sample desaturation. Similar to the Bond number N_B, the capillary number N_C describes the relation of the viscous forces and interfacial tension of two immiscible fluids [72]

$$N_C = \frac{v \cdot \mu}{\sigma}. \qquad (7.2)$$

Inserting the flow in velocity v in $[m/s]$, the viscosity μ in $[Pas]$ and the interfacial tension σ in $[N/m]$ into Equation 7.2, the capillary number becomes a dimensionless quantity. To avoid sample desaturation, the corresponding capillary number of each injection rate was calculated. In line with Section 6.1 and the work of Humphry et al. [53], this study assumed a critical capillary number in the range of 10^{-4} to 10^{-3}. After testing the three injection brines at a field rate equivalent injection of $0.05 \, cm^3/min$, the implemented USS corefloodings included the brine injection at $0.2, 1, 3$ and $5 \, cm^3/min$. As summarized in Table 7.1, the highest injection rate of $5 \, cm^3/min$ corresponds to capillary number of approximately 10^{-5}. The critical number of non-water-wet carbonates is hence not exceeded [53].

Once the secondary brine injection at an injection rate of $5 \, cm^3/min$ was stabilized, the obtained differential pressure was used to calculate the effective water permeability at remaining oil saturation $kw_{So,rm}$. The effective water permeability at remaining oil saturation represents the opposite of the effective oil permeability at connate water saturation $ko_{Sw,c}$, which was calculated in Chapter 4. Both effective permeability values are crucial for the numerical simulations, in which the maximum oil kr_o^{max} and maximum water endpoint relative permeability kr_w^{max} definition significantly impacts the history matching

7.2 METHODOLOGY

$$kr_o^{max} = \frac{ko_{Sw,c}}{k}, \qquad (7.3)$$

$$kr_w^{max} = \frac{kw_{So,rm}}{k}. \qquad (7.4)$$

7.2 Methodology

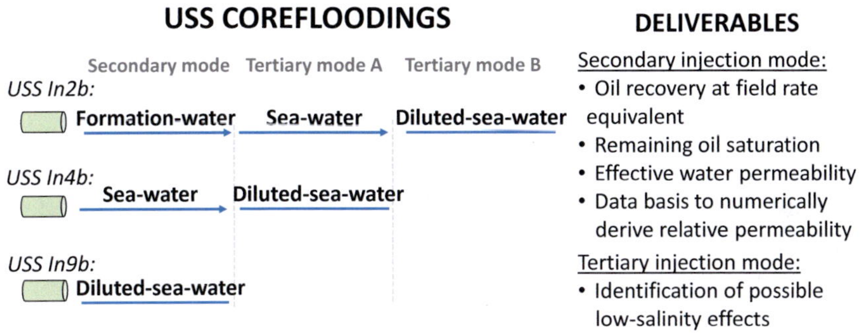

USS COREFLOODINGS

USS In2b: Secondary mode: Formation-water; Tertiary mode A: Sea-water; Tertiary mode B: Diluted-sea-water

USS In4b: Secondary mode: Sea-water; Tertiary mode A: Diluted-sea-water

USS In9b: Secondary mode: Diluted-sea-water

DELIVERABLES

Secondary injection mode:
- Oil recovery at field rate equivalent
- Remaining oil saturation
- Effective water permeability
- Data basis to numerically derive relative permeability

Tertiary injection mode:
- Identification of possible low-salinity effects

Figure 7.2: USS coreflooding methodology - The displacement efficiency of each brine was tested in secondary and tertiary injection mode application.

The experimental coreflooding methodology study is summarized in Figure 7.2. In regard to the promising results of the spontaneous imbibition and centrifuge method, the study included the conduction of three USS corefloodings. Injection of Formation-water in secondary mode followed by Sea-water and Diluted-sea-water in tertiary mode (1), injection of Sea-water in secondary mode followed by Diluted-sea-water in tertiary mode (2), and finally, the injection of Diluted-sea-water in secondary mode (3). While the secondary injection mode represents the injection of brine at connate water saturation, the tertiary injection mode expresses the exchange of the injection brine at a mature recovery stage. Thereby, all core samples contained Formation-water as connate water.

7.3 EXPERIMENTAL CONDUCTION

Initially, the selected brine was injected at a field rate equivalent of $0.05\, cm^3/min$ followed by a rate bump of $0.2, 1, 3$ and $5\, cm^3/min$. Depending on the secondary injection brine, the secondary mode was followed by the tertiary injection of a lower saline brine.

The secondary mode injection allowed the displacement efficiency comparison of Formation-water, Sea-water and Diluted-sea-water. Besides the oil recovery at field rate equivalent, the study investigated a possible brine composition impact on reaming oil saturation. Furthermore, the secondary injection mode differential pressure and water saturation data were used to derive relative permeability curves numerically.

After completing the secondary brine injection and determining the remaining oil saturation, corefloodings In2b and In4b included the exchange of the injection brine. Thereby, the tertiary injection of Sea-water and Diluted-sea-water evaluated possible Enhanced oil recovery effects. The applied tertiary injection rates of $0.05, 0.2, 1, 3$ and $5\, cm^3/min$ were thereby in line with the secondary mode injection rates. A similar coreflooding design was, for example, presented in the work of Nasralla et al. [82].

7.3 Experimental conduction

A Vinci autoflood 700 was used for the experiments. The device ensures coreflooding experiments under a maximum operating pressure of $700\, bar$ and $150\,°C$. The schematics of the coreflooding device is plotted in Figure 7.3.

The centerpiece of the coreflooding equipment is represented by a one-foot long core holder. After mounting the core sample inside a sleeve, the annulus between the sleeve and core holder was pressurized by silicon oil. Besides avoiding fluid flow between sleeve and core, the confining pressure simulates the overburden pressure. In the case of the three conducted USS experiments, the confining pressure was set to $103\, bar$ ($1500\, psi$).

The standard Vinci Autoflood version includes three fluid accumulators, which allowed the simultaneous loading of Formation-

7.3 EXPERIMENTAL CONDUCTION

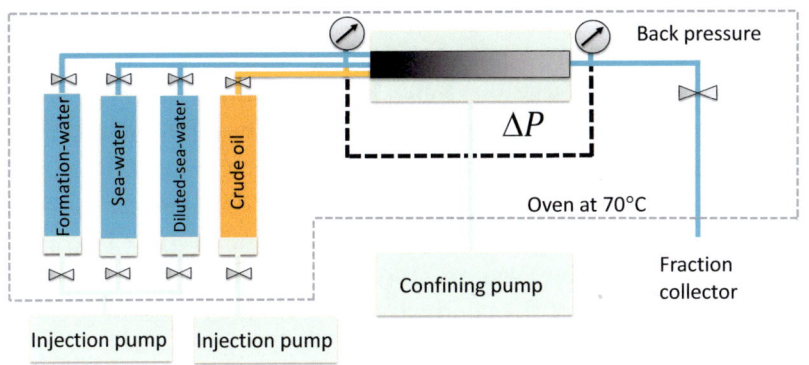

Figure 7.3: Experimental coreflooding sketch - Formation-water, Sea-water, Diluted-sea-water and crude oil are connected to the core holder. The oil accumulator is operated by an independent pump to ensure oil injection during the heating up of the system. A back-pressure system controls the pore pressure while the effluent is collected by an automatized fraction collector.

water, Sea-water and Diluted-sea-water. To ensure crude oil injection, the system was furthermore extended by a fourth accumulator. Before connecting the different accumulators to the core holder, approximately $50\,cm^3$ of the added fluid was produced. Besides removing possible impurities from the injection lines, the arising silicon oil buffer below the accumulator piston allowed the pressure balance during the heating up of the system.

While reaching the targeted test temperature of 70°C, the pressure of the Formation-water, Sea-water and Diluted-sea-water was controlled in constant pump pressure mode. A second injection pump injected crude oil into the core sample to prevent the sample draining. Moreover, the continuous oil injection ensured stabilized flow conditions as the injection fluid was exchanged to brine. The heating up of the system required approximately $5\,hours$.

After the injection into the core sample, the injected fluid moved towards the production outlet. The emerging differential pressure between the injection and production side was recorded by an automatized pressure transducer system. Depending on the differential pressure, the Vinci Autoflood automatically selects between three pressure range sensitive transducers.

The pore pressure was controlled by a back-pressure valve, which is located between the core holder outlet and fraction collector (cf. Figure 7.3). To acquire reasonable differential pressure data, it is crucial to establish the pore pressure before the displacement process is initiated. Therefore, the injection of crude oil at the beginning of the experiment was used to step-wise build the targeted back-pressure. In the case of the conducted corefloodings, a pore-pressure of $14\,bar$ ($200\,psi$) was selected.

In regard to the duration of the coreflooding experiments (72 to 120 $hours$), an automatized fraction collector was connected to the outlet of the coreflooding device. In line with the spontaneous and forced imbibition tests, the coreflooding experiments were conducted at 70°C.

After completing the coreflooding experiments, the obtained oil production was validated against Dean-Stark extraction. Therefore, the samples were placed inside a Dean-Stark apparatus and exposed to vaporized toluene. Once the toluene displaces the water from the sample, the vaporized water is captured inside a cooling unit. After the condensation of the water, the amount of displaced water is determined with the help of a graduated measuring tube. Since toluene does not dissolve salts, the amount of captured water requires a salt concentration correction

$$S_{DS} = \frac{V_w}{V_P}, \qquad (7.5)$$

where Sw_{DS} is the Dean-Stark validated water saturation, V_w is the captured water volume in $[cm^3]$ and V_P is the pore volume in $[cm^3]$.

7.4 Coreflooding results

This section briefly describes the experimental results of coreflooding In2b, In4b and In9b. A detailed analysis of the experimental and

7.4 COREFLOODING RESULTS

numerical coreflooding data is provided in Chapter 9. Furthermore, a quantitative summary of the experimental coreflooding results is summarized in Table 7.1.

7.4.1 Formation-water in secondary mode

The pressure and water saturation profile of the injection of Formation-water in secondary injection mode and Sea-water and Diluted-sea-water in tertiary mode is plotted in Figure 7.4.

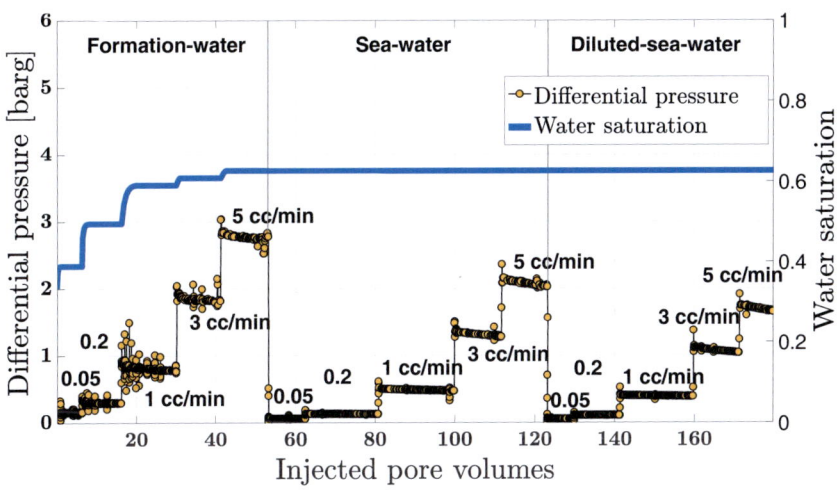

Figure 7.4: USS coreflooding In2b - Injection of Formation-water in secondary mode followed by the injection of Sea-water and Diluted-sea-water.

Injecting Formation-water at field rate equivalent caused an oil recovery of 21.9 %. The majority of the oil remained immobile, possibly due to heterogeneity, unstable displacement, capillary end effect and capillary trapping. The subsequent injection rates of $0.2, 1, 3$ and $5\,cm^3/min$ yielded into remaining oil saturation of 37.4 %. Due to the exchange of the injection brine, Figure 7.4 records a significant differential pressure decrease, which is explained by the injection brine viscosity reduction. Additional oil recovery due to the tertiary injection of Sea-water and Diluted-sea-water was not observed.

7.4 COREFLOODING RESULTS

7.4.2 Sea-water in secondary mode

The second coreflooding started with the injection of Sea-water in secondary mode followed by the injection of Diluted-sea-water in tertiary mode (Figure 7.5). At the first injection rate of $0.05\,cm^3/min$, 30.1 % of the oil was recovered. After bumping up the injection rates, a remaining oil saturation of 35.5 % was obtained. The following injection of Diluted-sea-water led to an additional oil recovery of 3.4 % at an injection rate of $1\,cm^3/min$, followed by additional 1.8 % oil recovery at an injection of $3\,cm^3/min$ and $5\,cm^3/min$. The oil saturation at the end of the experiment was 31.7 %.

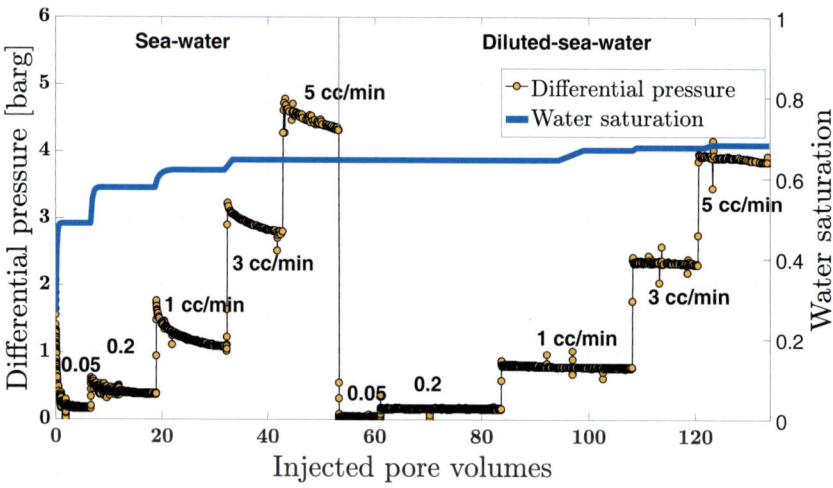

Figure 7.5: USS coreflooding In4b - Injection of Sea-water in secondary mode followed by the injection of Diluted-sea-water.

7.4.3 Diluted-sea-water in secondary mode

Figure 7.6 shows an oil recovery of 38.5 % for the injection of Diluted-sea-water at field rate. In comparison to the injection of Formation-water (coreflooding In2b) and the injection of Sea-water (coreflooding In4b) at field rate equivalent, coreflooding In9b resulted in the

7.5 SUMMARY & CONCLUSIONS

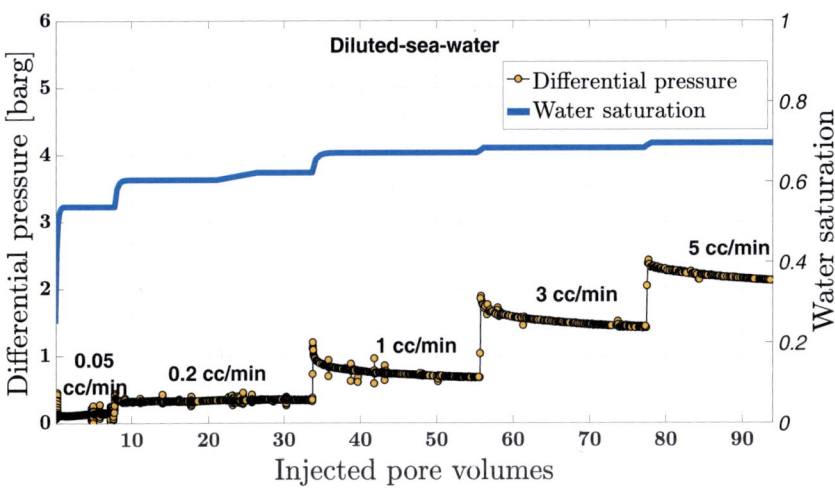

Figure 7.6: USS coreflooding In9b - Injection of Diluted-sea-water in secondary mode.

significantly highest oil recovery. Bumping up the injection rates led to a remaining oil saturation of 30.6 %.

7.5 Summary & conclusions

During the secondary injection mode, all three conducted corefloodings showed a similar saturation and pressure response. After the injection at field rate equivalent, the subsequent rate bumping of $0.2, 1, 3$ and $5\,cm^3/min$ caused a significant reduction of the remaining oil saturation. A similar production pattern was observed in the work of Nasralla et al. [82] in which unsteady state corefloodings on carbonate reservoir samples were conducted. The behavior indicates that the corefloodings were significantly impacted by end effects. An overview of the experimental coreflooding results is provided in Table 7.1.

The spontaneous and forced imbibition experiments showed a correlation between the system's salinity and the corresponding wettability, capillary pressure and residual oil saturation. In secondary

7.5 SUMMARY & CONCLUSIONS

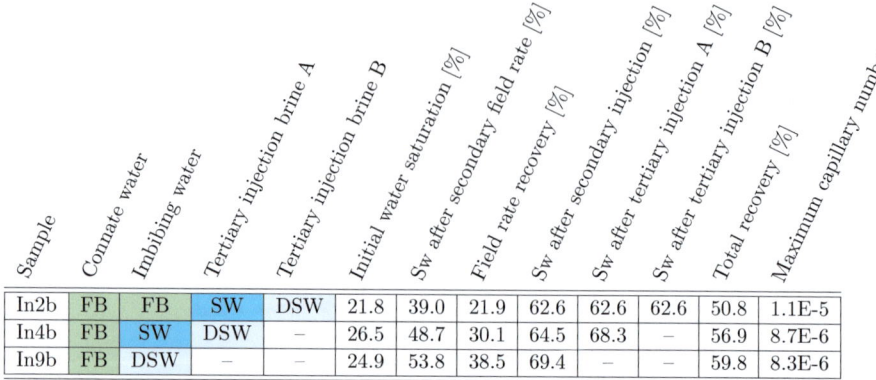

Sample	Connate water	Imbibing water	Tertiary injection brine A	Tertiary injection brine B	Initial water saturation [%]	Sw after secondary field rate [%]	Field rate recovery [%]	Sw after secondary injection [%]	Sw after tertiary injection A [%]	Sw after tertiary injection B [%]	Total recovery [%]	Maximum capillary number
In2b	FB	FB	SW	DSW	21.8	39.0	21.9	62.6	62.6	62.6	50.8	1.1E-5
In4b	FB	SW	DSW	–	26.5	48.7	30.1	64.5	68.3	–	56.9	8.7E-6
In9b	FB	DSW	–	–	24.9	53.8	38.5	69.4	–	–	59.8	8.3E-6

Table 7.1: USS coreflooding results - A detailed summary of the conducted USS coreflooding results In2b, In4b and In9b.

injection mode, Diluted-sea-water caused the highest oil recovery at field rate equivalent as well as after rate bumping. The potential of Diluted-sea-water to increase oil recovery in tertiary remains questionable, as additional oil recovery was only observed in one out of two experiments.

Numerical history matching is a crucial step during the processing and interpretation of unsteady state coreflooding data. Therefore, the three conducted USS displacement experiments are discussed in more detail in Chapter 9.

- This chapter describes the fundamentals of the unsteady state coreflooding technique. Besides illustrating the origin of capillary end effects during unsteady state corefloodings, the chapter furthermore introduces the concept of rate bumping to overcome end effects.

- Three unsteady state corefloodings were successfully conducted. Depending on the coreflooding design, 90 to 180 pore volume equivalents were injected at a temperature of 70°C.

- Applying an injection rate of $0.05\ cm^3/min$ in secondary injection mode, Formation-water caused 21.9 % oil recovery, Sea-

7.5 SUMMARY & CONCLUSIONS

water caused 30.1 % oil recovery and Diluted-sea-water caused 38.5 % oil recovery. While this is not representative of reservoir conditions due to capillary end effects, this demonstrates the impact of injection water on wettability and potentially relative permeability.

- When comparing the secondary mode experiments, the final remaining oil saturation for Formation-water injection, Sea-water injection and Diluted-sea-water injection was 37.4 %, 35.5 % and 30.6 %, respectively. Minor additional oil recovery occurred when injecting Diluted-sea-water in tertiary injection mode in one out of two corefloodings.

Part II
Numerical study

8
Numerical centrifuge simulation

The characteristic physics of the centrifuge method requires an analytical and preferably numerical analysis of the experimentally obtained data. Specialized SCAL software such as Cydar and Sendra provide numerical tools which allow the simulation of the centrifuge method [60]. Furthermore, CMG IMEX contains a simulation tutorial of a centrifuge experiment of Maas et al. [62].

Commercial software solutions reveal limited information on the mathematical and numerical implementation. To ensure the full understanding and control of the numerical simulations, this thesis therefore aims to develop an independent numerical centrifuge and coreflooding model. In order to avoid a tedious model development from scratch, the simulation was built on top of the open-source simulator DuMu^x. The developed numerical centrifuge and coreflooding models were verified against the commercial Cydar software.

8.1 DuMu^x

DuMu^x is an open-source simulator for continuum-scale fluid flow in porous media. Licensed under the GNU General Public License (GPL), the C++ source code can be fully accessed, modified and ex-

8.1 DUMU$^\text{x}$

tended by the user. DuMu$^\text{x}$ offers a large variety of tutorials, starting from simple one-phase fluid flow to compositional three-phase gas-water-oil cases. Isothermal, as well as non-isothermal models, are implemented. Besides flow models, DuMu$^\text{x}$ provides a comprehensive collection of pre-programmed components, fluid systems, and fluid matrix interaction models. However, since the majority of the available material focuses on gas-water systems, petroleum engineering related problems require a comprehensive adaption. The partial differential equations are solved with grid-based methods by using the Distributed and Unified Numeric Environment (DUNE) [36].

DuMu$^\text{x}$ provides a fully-implicit as well as a semi-implicit numerical scheme. The fully-implicit approach supports box and cell-centered spatial discretization schemes while the time discretization is implemented by the Backward Euler method. Since both spatial discretization approaches have advantages and disadvantages, the discretization scheme needs to be selected in accordance with the simulation specific requirements. As a primary benefit of the cell-centered discretization scheme, the method is assumed to be mass conservative. While the cell-centered method uses the finite volume method to process structured grid geometries, the box method allows the processing of unstructured grids by combining the finite volume method and finite element method [36] [32].

8.1.1 Spatial discretization

The open-source simulator DuMu$^\text{x}$ and the commercial SCAL simulator Cydar were used to history match the centrifuge and coreflooding experiments numerically. The implemented simulations are based on a fully-implicit numerical scheme.

In regard to the rectangular core grids and time efficiency, the DuMu$^\text{x}$ centrifuge and coreflooding models are using the mass conservative cell-centered finite volume method as spatial discretization scheme. In general, the finite volume method introduces a (finite) number of non-overlapping control volumes (cell blocks) to discretize a problem spatially. The properties of the cells are thereby described by the cell center (cf. Figure 8.1) [50].

8.1 DUMU$^\text{X}$

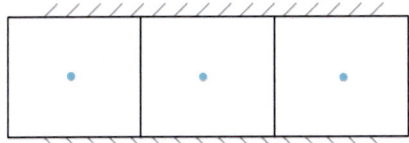

Figure 8.1: Cell-centered finite volume method - The DuMu$^\text{x}$ cell-centered finite volume method is limited to the processing of unstructured grids. The properties of the non-overlapping cells are described by the cell centers.

The two-phase-two-component mass balance of the centrifuge model can be summarized by Equation 8.1 (cf. Section 8.2)

$$\phi \frac{\partial (\rho_o S_o + \rho_w S_w)}{\partial t} + \nabla \cdot (\rho_w \, v_w + \rho_o v_o) = q_i, \quad i = o, w, \quad (8.1)$$

where ϕ is the porosity, ρ_i is the fluid density in $[kg/m^3]$, S_i is the fluid saturation, v_i is the advective volumetric velocity in $[m/s]$ and q_i is the source/sink term $[kg/(s \cdot m^3)]$. The subscripts o and w represent and oil and water phase, respectively. The mass balance can be rewritten in a more generalized form [32]

$$\underbrace{\frac{\partial}{\partial t} u}_{1} + \underbrace{\nabla \cdot f(u)}_{2} = \underbrace{q(u)}_{3}, \quad (8.2)$$

where the time-depending change of the depending variables is represented by the first term on the left-hand (1) and the second term (2) expresses the grid in- and outflow. While the flux term of the two-phase centrifuge model is limited to consideration of the advective flux, the two-phase-three-component coreflooding model additionally considers the impact of the component diffusion. The right-hand side term (3) denotes the source and sink terms [36, 47, 88]. In order to combine the spatial and temporal discretization, Equation 8.2 requires some fundamental conversions. Initially, Equation 8.2 is integrated over each control volume [88]

8.1 DUMU$^\text{X}$

$$\underbrace{\int_V \frac{\partial u}{\partial t} dV}_{1} + \underbrace{\int_V \nabla \cdot f(u) dV}_{2} + \underbrace{\int_V q(u) dV}_{3} = 0. \tag{8.3}$$

Since the control volumes (grids) are itself independent of time, the differential and integral of the time-depending term (1) can be exchanged. Furthermore, the application of Gauss's theorem allows the conversion from volume to surface integral [88]

$$\frac{d}{dt} \int_{V_i} u dV_i + \int_{\partial A_i} f(u) \cdot n dA_i + \int_{V_i} q(u) dV_i = 0. \tag{8.4}$$

The obtained Equation 8.4 is a regular differential equation which depends on the in- and outflux $f(u)$ and the source/sink term $q(u)$ of the control volumes. The application of a spatial discretization method allows the introduction of average values to replace the integrals. The time depending term u of any grid i can be thereby described as following [47, 50]

$$u_i = \frac{1}{V_i} \int_{V_i} u dV_i, \tag{8.5}$$

and similar for the source/sink term

$$q_i = \frac{1}{V_i} \int_{V_i} q(u) dV_i. \tag{8.6}$$

The integral of the flux term is replaced by a numerical scheme. DuMu$^\text{x}$ processes Two-point flux approximation (TPFA) as well as Multi-point flux approximation (MPFA) schemes [32]. Equation 8.7 uses a quadrature approximation to approximate the flux calculation

$$\int_{\partial A_i} f(u) \cdot n dA \approx \sum_{j=1}^{n} k_{i,j}(u_i, u_j), \tag{8.7}$$

8.1 DUMU$^\text{X}$

where term $k_{i,j}$ denotes the geometrical grid, rock, and fluid properties [47]. The total amount of grids is represented by n and the subscripts i and j are representing the neighboring grid cells. Equation 8.4 is finally rewritten by substituting the integrals by Equation 8.5, Equation 8.6 and Equation 8.7 [47, 50]

$$\frac{d}{dt}u_i + \frac{1}{V_i}\sum_{j=1}^{n} k_{i,j}(u_i, u_j) + q(u_i) = 0. \tag{8.8}$$

Equation 8.8 is representative of the spatial discretization scheme of the centrifuge and coreflooding model. The next section completes the introduction of the fully-implicit numerical scheme by coupling the spatial discretization and the temporal discretization.

8.1.2 Time discretization

The Backward Euler method is an approximation method for a first-order ordinary differential equation. A general form of a first-order differential equation can be written as following [13] [43]

$$\frac{dx(t)}{dt} = f(t, x(t)). \tag{8.9}$$

The Backward Euler approximation method assumes that x_t at any time step equals x_{t-1} at the previous time step. In other words, the slope is determined by looking (backward) to the previous time step

$$\frac{dx(t)}{dt} \approx \frac{x_t - x_{t-1}}{\Delta t}. \tag{8.10}$$

Since numerical simulators are used to forecast the behavior of a system, the indexing of Equation 8.10 might be misleading. By convention, the index notion of Equation 8.10 is modified

$$\frac{dx(t)}{dt} \approx \frac{x_{t+1} - x_t}{\Delta t}. \tag{8.11}$$

8.1 DUMU^X

In regard to the notation of the generalized material balance (Equation 8.2), Equation 8.11 might be adapted as following

$$\frac{du(t)}{dt} \approx \frac{u^{t+1} - u^t}{\Delta t}. \tag{8.12}$$

Inserting the Backward Euler method approximation into the spatially discretized mass equation (Equation 8.8) leads to the fully-implicit solution of the mass balance

$$u_i^{t+1} = u_i^t - \frac{\Delta t}{V_i} \sum_{j=1}^{n} k_{ij}(u_i^{t+1}, u_j^{t+1}) + q(u_i^{t+1}) = 0. \tag{8.13}$$

8.1.3 Newton's method

Equation 8.13 couples the spatial and temporal discretization to the mass balance. The coupled discretization results in a matrix that needs to be solved at each time step. DuMu^x as well as Cydar are using Newton's method to assemble a linear solving system. In general, Newton's method (or Newton–Raphson method) is an iterative approach to find the roots (zeros) of a function

$$x_{n+1} = x_n - \frac{f(x_n)}{f'(x_n)}. \tag{8.14}$$

The power of Newton's method can be demonstrated by a simple example. Instead of directly calculating the root of $\sqrt{16}$, Newton's method consecutively approximates the result. In order to apply Newton's method to determine an approximation of $\sqrt{16}$, the problem is initially rewritten as a non-linear differential equation

$$f(x_n) = \sqrt{16} = x_n, \quad \Leftrightarrow \quad = x_n^2 - 16 = 0. \tag{8.15}$$

The derivative of Equation 8.15 is straightforward

8.1 DUMU^X

$$f'(x_n) = 2 \cdot x_n = 0. \tag{8.16}$$

The first iterative of Newton's method (Equation 8.14) is determined by calculating $f(x_n)$ and $f'(x)$ based on a first guess $x_n = 3$

$$x_{n+1} = 3 - \frac{3^2 - 16}{2 \cdot 3} = 4.167. \tag{8.17}$$

Equation 8.17 shows that despite the fact of an unrealistic first guess of $x_n = 3$, the first iterative yields into a reasonably close approximation of $\sqrt{16}$. The results of the subsequent iterations are summarized in Table 8.1. The iterative method consecutively increases the accuracy of the approximation. After four iterations, the precision of approximation amounts already 13 decimal places.

Iterative	x_n	Approximation
1	3	4.167
2	4.167	4.0034
3	4.0034	4.0000014
4	4.0000014	4.00000000000024

Table 8.1: Newton's method - The example illustrates the consecutive approximation of $\sqrt{16}$.

In the case of numerical simulation software, the non-linear vector notation complicates the application of Newton's method. Since the determination of the derivative $f'(x)$ is less straight forward and a division of two vectors is not possible, the DuMu^x Newton uses a slightly different approach. In the first place, the Newton solver sums up the time-depending term (1), the grid in- and outflow (2) and the source/sink term (3) of Equation 8.18 to assemble the Residual matrix $r(u)$ [33]

$$\underbrace{\frac{\partial}{\partial t} u + \nabla \cdot f(u) + q(u)}_{r(u)} = 0. \tag{8.18}$$

8.2 DUMU^X CENTRIFUGE SIMULATION

Once the residual matrix is constructed, the first-order partial derivative of $r(u)$ are assembled inside the Jacobian Matrix J_r

$$\frac{d}{du}r(u) = J_r(u). \tag{8.19}$$

Based on a first guess u_n, the values of the Residual and Jacobian matrix are calculated. Instead of dividing the residual vector by the Jacobian matrix, the linear variable x_n is introduced to denote the difference between $J_r(u_n)$ and $r(u_n)$

$$J_r u_n \cdot x_n = r(u_n). \tag{8.20}$$

The divergence between the two vectors is reduced at each iteration by subtracting x_n from the previous guess u_n

$$u_{n+1} = u_n - x_n. \tag{8.21}$$

The Newton solver repeats the iteration scheme until a defined minimum divergence is obtained [33].

8.2 DuMu^x centrifuge simulation

After describing the fully-implicit numerical scheme of the DuMu^x software, this section introduces the mathematical model and the numerical implementation of the centrifuge method. Source code excerpts are provided in Appendix B.

8.2.1 Boundary conditions

The numerical representation of a physical process typically starts with a careful selection of reasonable boundary conditions. DuMu^x processes boundary conditions of the first (Dirichlet boundary) and second kind (Neumann boundary). While the two-phase Dirichlet boundary condition is specified by a static pressure and saturation

8.2 DUMU^X CENTRIFUGE SIMULATION

value, the Neumann boundary condition requires the definition of a source/sink term [24]. Figure 8.2 schematically summarizes the imbibition centrifuge model, in which Dirichlet boundary conditions define the inlet and outlet face fluid flow.

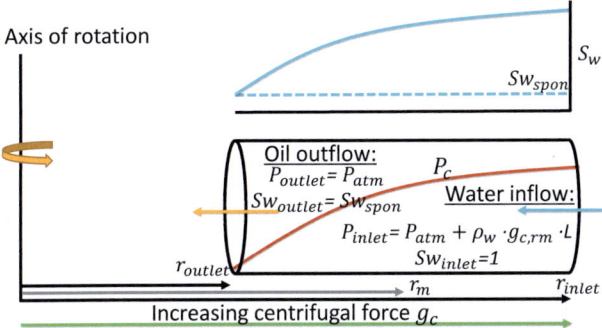

Figure 8.2: Numerical centrifuge imbibition model - The arising centrifugal force causes the oil mobilization towards the center of rotation. The produced oil is replaced by imbibing water. The in- and outflow calculation is imposed by Dirichlet boundary conditions.

After completing the spontaneous imbibition, the core samples were placed inside the centrifuge and surrounded by imbibing brine. At this point, the water saturation of the core corresponded to Sw_{spon}. As illustrated in Figure 8.2, the arising centrifugal acceleration of the centrifuge technique causes oil flow towards the center of rotation. As a result of the oil-water density difference, the oil exits the core at the outlet face (cf. Figure 8.2, r_{outlet}). The mobilized oil is replaced by the imbibing brine, which enters the core at the inlet face (Figure 8.2, r_{inlet}).

The mathematical formulation of the inlet and outlet fluid flow is summarized by Equation 8.23 and Equation 8.24 [63]. In the case of the inlet pressure boundary condition, initially, the average centrifugal acceleration $g_{c,rm}$ along the core is calculated

8.2 DUMU^X CENTRIFUGE SIMULATION

$$g_{c,rm} = \left(\frac{RPM \cdot 2\pi}{60}\right)^2 \cdot r_m, \qquad (8.22)$$

where RPM represents the centrifuge velocity in revolutions per minute and r_m is the distance of the center of rotation to the center of core in $[m]$. The inlet pressure P_{inlet} is then calculated based on the atmospheric P_{atm} in $[Pa]$, the water density ρ_w in $[kg/m^3]$ and the core length L in $[m]$

$$P_{inlet} = P_{atm} + \rho_w \cdot g_{c,rm} \cdot L, \qquad \text{where } Sw_{inlet} = 1. \qquad (8.23)$$

The Dirichlet inlet face boundary condition formulation is completed by defining a constant inlet water saturation of 100 %. This becomes evident, as only water accumulates at the highest point of centrifugal acceleration.

Due to the centrifugal forces, the mobilized oil moves towards the center of rotation. Hassler & Brunner first formulated, that the saturation of the wetting fluid at the outlet remains at a saturation of 100 % [49]. However, in order to take the conducted spontaneous imbibition experiments into account, the presented numerical model assumes a constant outlet water saturation of Sw_{spon}. Furthermore, water does not break through and capillary pressure is set to zero. Consequently, the outlet face Dirichlet condition reduces to Equation 8.24

$$P_{outlet} = P_{atm}, \qquad \text{where } Sw_{outlet} = Sw_{spon}, \qquad (8.24)$$

where P_{outlet} is the outlet pressure in $[Pa]$. The formulation of the boundary conditions is completed by defining Neumann no-flow boundaries along the core. The source code implementation of the Dirichlet boundary conditions is illustrated in Appendix B.1 while the centrifugal force implementation is displayed in Appendix B.2.

8.2 DUMU^X CENTRIFUGE SIMULATION

8.2.2 Fluid properties

The centrifuge model is built on top of an immiscible two-phase oil-water fluid system. The phases are assumed to be in thermodynamic equilibrium at a temperature of 70°C. Furthermore, the pressure impact on fluid properties is assumed to be negligible. In contrast to the compositional coreflooding system, the oil and water phase proprieties are directly specified for each phase. The fluid densities and viscosities are defined as measured during the experimental study (cf. Chapter 4)

$$\rho_i = \rho_{i,measured}, \quad i = o, w, \quad (8.25)$$

and

$$\mu_i = \mu_{i,measured}, \quad i = o, w, \quad (8.26)$$

where ρ_i is the density in $[kg/m^3]$, μ_i is the viscosity in $[Pas]$ and the subscripts o and w denote the oil and water phase.

8.2.3 Hydraulic properties

The numerical centrifuge model defines capillary pressure as the differential pressure between the immiscible oil and water phase. In other words, the capillary pressure P_c is the resistance that needs to be overcome to initiate flow between the oil and water phase

$$P_c(S_w) = P_o - P_w, \quad (8.27)$$

where P_o is the oil phase pressure in $[Pa]$ and P_w is the water phase pressure in $[Pa]$.

The shape of a typical capillary pressure drainage curve is illustrated in Figure 8.3, a. After overcoming the entry pressure, the capillary pressure inside an initially 100 % wetting fluid saturated

8.2 DUMU^X CENTRIFUGE SIMULATION

porous media increases with decreasing wetting fluid saturation [2]. While the smaller capillary pressure inside the larger pores allows an efficient and immediate wetting fluid drainage, the higher capillary pressure of smaller pores mitigates the drainage process. Once the wetting fluid reaches the residual saturation, the capillary pressure significantly increases and hence traps the wetting fluid.

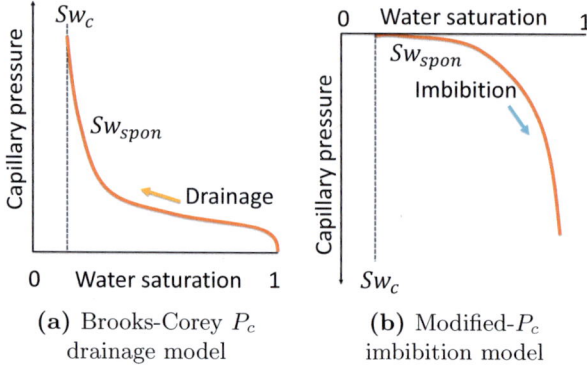

(a) Brooks-Corey P_c drainage model

(b) Modified-P_c imbibition model

Figure 8.3: Capillary pressure model comparison - The default Brooks-Corey P_c drainage model does not reflect the physics of an imbibition process. It was therefore replaced by a Modified-P_c model.

DuMu^x provides several default models that are representative of the typical capillary pressure drainage behavior. However, in regard to the conducted imbibition experiments, a modification of the default capillary pressure models is required. Reminding the initial water saturation of Sw_{spon} and considering Figure 8.3, a, it becomes obvious, that the application of a drainage capillary pressure model results into physically incorrect imbibition simulation. In this case, the capillary pressure at the beginning of the experiment is characterized by a pressure peak and subsequently decreases with increasing water saturation. This is, however, not representative of a forced water-oil imbibition experiment, in which capillary pressure increases with increasing water saturation (Figure 8.3, b,). To represent this characteristic forced imbibition behavior, a Modified-P_c law was implemented inside DuMu^x. The least-square solution of

8.2 DUMU$^\text{X}$ CENTRIFUGE SIMULATION

a hyperbolic function thereby replaces the standard Brooks-Corey capillary pressure model

$$P_c = \frac{B + C \cdot Sw_e}{1 + D \cdot Sw_e}, \tag{8.28}$$

where P_c is the capillary pressure in $[Pa]$ and the parameters B, C, D are the least square solutions of the hyperbola. A similar mathematical approach was used for the analytical centrifuge solution, where the hyperbolic function regresses the experimentally obtained centrifuge data [30] (cf. Equation 6.10). The parameter Sw_e denotes the effective or normalized water saturation

$$S_{we} = \frac{Sw - Sw_c}{1 - Sw_c - So_r}, \tag{8.29}$$

where Sw_c is the connate water saturation and So_r is the residual oil saturation.

Besides capillary pressure as a hydraulic property, the formulation of the relative permeability model significantly impacts the results of the centrifuge simulation. Since the DuMu$^\text{x}$ default Brooks-Corey relative permeability model only allows the adjustment of a single input parameter, it is difficult to obtain a satisfying history match. The default model was therefore exchanged by a Modified-kr model to improve the input relative permeability options.

The Modified-kr model significantly extends the control of the relative permeability input curves by introducing three parameters into model. During the numerical simulation, each variable can be adjusted to improve the obtained history match. In comparison to the Brooks-Corey model, especially the control of the curve shape around the origin and end of the relative permeability curves is improved.

To ensure the comparability of the DuMu$^\text{x}$ and Cydar simulations, the mathematical formulation of the Modified-kr formulation was taken from the Cydar simulator [27]. The Modified-kr model

8.2 DUMU^X CENTRIFUGE SIMULATION

parameters ϵ, H and V are directly entered by the user and replace the traditional Corey exponent

$$kr_w(S_w) = kr_w^{max} \cdot \left(\frac{e}{2\epsilon} \cdot Sw_e^{2\epsilon} + \frac{f}{\epsilon} \cdot Sw_e^e \cdot Sw_e^H\right), \qquad (8.30)$$

and for the oil phase

$$kr_o(Sw) = kr_o^{max} \cdot \left(\frac{e}{2\epsilon} \cdot (1 - Sw_e)^{2\epsilon} + \frac{f}{\epsilon} \cdot (1 - Sw_e)^e \cdot (1 - Sw_e)^H\right), \quad (8.31)$$

where kr_w^{max} is the water endpoint relative permeability and kr_o^{max} is the oil endpoint relative permeability. The quantities of kr_w^{max} and kr_o^{max} are calculated based on the effective endpoint and absolute permeabilities (cf. Chapter 7, Equation 7.3 and Equation 7.4). The terms e and d are representing following quantities

$$e = G - H - b, \qquad (8.32)$$

$$f = 2\epsilon(1 - H) - G - H. \qquad (8.33)$$

The source code implementations of the Modified-P_c and Modified-kr model are listed in Appendix B.3.

8.2.4 Flow equation

The flow model uses a standard multi-phase Darcy approach as the equation of the conservation of momentum. The gravitational acceleration is thereby replaced by the centrifugal acceleration g_c (cf. Equation 8.35)

$$v_i = -\frac{k \cdot kr_i}{\mu_i} \cdot (\nabla P_i - \rho_i g_c), \quad i = o, w, \qquad (8.34)$$

8.2 DUMU^X CENTRIFUGE SIMULATION

where v_i is the advective volumetric velocity in $[m/s]$, k is the absolute permeability in $[m^2]$, kr_i is the relative permeability, μ_i is the fluid viscosity in $[Pas]$ and P_i is the phase pressure in $[Pa]$, ρ_i is the phase density in $[kg/m^3]$ and the subscripts o and w represent the oil and water phase.

The centrifugal acceleration is the driving force during the centrifuge experiments and hence requires an accurate numerical implementation. The apparent centrifugal acceleration depends on the applied centrifuge spin RPM and increases from the center of rotation radially outwards from the origin (cf. Figure 8.2). In the implemented numerical model, the centrifugal acceleration g_c is calculated for each grid node and thus varies along the core sample

$$g_c(r) = \omega^2 \cdot r = \left(\frac{RPM \cdot 2\pi}{60}\right)^2 \cdot r, \qquad (8.35)$$

where ω is the angular velocity in $[rad/s]$, r is the distance of the center of rotation to the node $[m]$, and RPM denotes the centrifuge velocity in revolutions per minute.

In case the centrifuge spin is changed, Equation 8.35 causes an immediate centrifugal acceleration increase. This is not representative of the physical reality, in which the centrifuge acceleration requires an equipment specific time duration. In order to mimic and buffer the centrifuge acceleration during the spin change, a build-up time is introduced into Equation 8.35

$$g_c(r) = \frac{\sum_{n=0}^{N} \Delta t_{BuildUp}}{t_{BuildUp}} \cdot \left(\frac{RPM \cdot 2\pi}{60}\right)^2 \cdot r, \qquad (8.36)$$

where $\Delta t_{BuildUp}$ is the time step size during the centrifuge acceleration in $[s]$ and $t_{BuildUp}$ is the total duration of the centrifuge acceleration in $[s]$. The presented centrifuge simulation assumes a total centrifuge acceleration duration $t_{BuildUp}$ of $100\,seconds$.

8.2 DUMU^X CENTRIFUGE SIMULATION

8.2.5 Mass balance

The advective flux (and hence the centrifugal force, Equation 8.35) is coupled to the mass balance as summarized in Equation 8.37

$$\phi \frac{\partial (\rho_o S_o + \rho_w S_w)}{\partial t} + \nabla \cdot (\rho_w\, v_w + \rho_o v_o) = q_i, \quad i = o, w, \quad (8.37)$$

where porosity is represented by ϕ and q_i is the oil and water phase source/sink term in $[kg/(s \cdot m^3)]$. The system of equations is closed by the sum of saturations

$$S_o + S_w = 1. \quad (8.38)$$

By combing the constitutive relations for the capillary pressure (Equation 8.27), relative permeability (Equation 8.30 and Equation 8.31) and using the closure Equation 8.38, the number of unknowns is reduced to two. As elaborated in Section 8.1.1 to Section 8.1.3, the material balance is discretized by using a cell-centered finite volume scheme as spatial and the implicit Backward Euler method as time discretization. The system of equations is solved by Newton's method [32].

8.2.6 Time manager

The implemented centrifuge model uses an episode manager to control the boundary conditions as a function of time. After the problem initialization, the simulation is consisting of eighteen consecutive time episodes. Each of the applied nine centrifuge spin steps is thereby controlled by two episodes. During the first 100 *seconds* of each centrifuge spin, the centrifugal acceleration is buffered by a steady increase of the spin velocity. Once the targeted centrifuge spin is reached, the second episode of each time step manages the time control of the remaining spin duration.

8.3 Cydar

The physical principle, boundary conditions and mathematical coupling of the DuMu^x centrifuge simulation was summarized in the previous section. To validate the developed model, the commercially established Cydar software was selected. The SCAL specialized software Cydar provides good comparability to DuMu^x, as both simulators are using a fully-implicit numerical scheme. Initially developed by the french IFP Energies nouvelles (IFPEN), Cydar was outsourced into the Cydarex company. The software provides several numerical tools to simulate a large variety of SCAL experiments, such as centrifuge method, unsteady state and steady state displacement experiments [60].

The implemented DuMu^x and Cydar simulations are based on an incompressible fully-implicit numerical scheme. The time depending term is thereby solved by the Backward Euler method, while the finite difference method is used as spatial discretization. Furthermore, Cydar is using the Newton-Raphson method to solve the system of equations [27].

8.4 Cydar centrifuge simulation

8.4.1 Cydar input data

In order to ensure the comparability of the software solutions, the implemented DuMu^x and Cydar simulations are based on the identical input data. Besides the oil and water density and viscosity at 70°C, Cydar requires the core and centrifuge dimension specifications to calculate the arising inlet pressure. Porosity, permeability and initial water saturation are assumed to be homogeneously distributed along the core sample. Furthermore, the applied centrifuge spins and duration are specified as applied during the experimental study.

The definition of hydraulic proprieties is crucial to obtain a reasonable history match. Cydar allows the relative permeability data

8.4 CYDAR CENTRIFUGE SIMULATION

input in the form of a Brooks-Corey, Modified-kr and LET model. Furthermore, relative permeability data can be entered as a saturation versus relative permeability input table. In line with the DuMu^x simulation, the Modified-kr model was selected to specify the relative permeabilities (cf. Equation 8.30 and Equation 8.31).

The initial capillary pressure data were defined as calculated by the analytical Forbes First solution (Equation 6.7).

8.4.2 Cydar boundary conditions

Due to the commercialization of Cydar, the information on the mathematical formulation and implementation of the centrifuge model is limited. In case of the imbibition centrifuge experiments, Cydar calculates the capillary inlet pressure P_{ci} as following [18, 27, 60]

$$P_{ci} = \frac{1}{2} \cdot \Delta \rho \cdot \omega^2 \cdot \left(r_{outlet}^2 - r_{inlet}^2 \right), \quad (8.39)$$

where $\Delta \rho$ is the density difference between the oil and water phase, ω is the angular velocity of the centrifuge, r_{outlet} is the distance of the core outlet face to the center of rotation and r_{inlet} is the distance of the core inlet face to the center of rotation.

At a first glance, the imposed inlet boundary appears to deviate from the implemented DuMu^x inlet boundary condition. (cf. Equation 8.23). However, while the DuMu^x boundary condition refers to the hydrostatic water pressure, Equation 8.39 directly determines the inlet capillary pressure. Since capillary pressure is defined as the differential pressure between the immiscible phases, the imposed DuMu^x (cf. Equation 8.23) and Cydar (cf. Equation 8.39) boundaries are leading in the identical inflow conditions.

In the case of the outlet boundary conditions, Cydar restricts the effluent production to the oil phase. In line with the DuMu^x model, the capillary outlet pressure is set to zero [60].

The work of Lenormand [60] described the principle of an incompressible unsteady state brine imbibition Cydar coreflooding model as following. Once the coreflooding simulation is started, the outlet

8.5 SIMULATION RESULTS

boundary condition initially allows the production of oil. Thereby, the amount of produced oil is identical to the amount of injected brine. After water breakthrough at the outlet face, the capillary outlet pressure remains at zero and the water and oil phase are co-produced.

8.5 Simulation results

The presented DuMu^x centrifuge model was verified against the commercial Cydar software by simulating the centrifuge experiment of sample In17. A 2D Cartesian 100 x 1 grid was selected to represent the core, in which the porosity and permeability were assumed to be homogeneously distributed. The numerical DuMu^x centrifuge model is plotted in Figure 8.6.

8.5.1 History matching methodology

The history matching followed a unified procedure. Initially, an intermediate-wet relative permeability system was defined by specifying the Modified-kr model parameters as $\epsilon = 2$, $H = 0$ and $G = 2$. The resulting relative permeability curves are identical to the curve shape of the traditional Brooks-Corey model when using a Corey exponent of 2.

The definition of the initial hydraulic properties was completed by specifying the capillary pressure as determined by Forbes first solution (cf. Equation 6.16).

After completing a first numerical simulation based on the initial input data, the comparison of the experimental and numerical saturation already resulted in a close saturation match (cf. Figure 8.4, red line). In the case of the centrifuge method, the resulting average water saturation at each centrifuge spin is predominantly impacted by capillary pressure. Consequently, the good agreement between the experimental and analytically corrected input data confirms the validity of the applied of the hyperbolic regression and Forbes first solution combination. However, a small adjustment of the input capillary pressure helped to improve the saturation match. While a

8.5 SIMULATION RESULTS

decrease of the capillary pressure (less negative) increases the water saturation, the capillary pressure increase caused a water saturation reduction.

In contrast to the coreflooding method, the adjustment of the relative permeability hardly impacts the numerically obtained water saturation. While the capillary pressure defines the final water saturation at each centrifuge spin, the relative permeability adjustment either delays or accelerates the oil production. Figure 8.4, red line shows the initial history match, in which the relative permeability is described by a neutral relative permeability of $\epsilon = 2$, $H = 0$ and $G = 2$. After adapting the input relative permeability, the relative permeability delays the oil production at each centrifuge speed (cf. Figure 8.4, blue line). While some studies are using the centrifuge method to obtain relative permeability of the effluent fluid [72], this work conducted unsteady corefloodings to derive relative permeability data.

Figure 8.5 shows, that the Cydar as well as the presented DuMu^x model result into an accurate match of the experimentally data. Note, that the plotted simulation results of Figure 8.5 are based on the identical fluid properties, core properties, capillary pressure and relative permeability data. The small saturation differences between the DuMu^x and Cydar history matches are likely to be caused by the mathematical description of the capillary pressure. While the implemented DuMu^x model uses a hyperbolic function to describe the capillary pressure, the Cydar simulation is based on a saturation versus capillary pressure input table.

8.5.2 Forced imbibition simulation

The experimental centrifuge data as well as the DuMu^x and Cydar history matches of sample In17 are plotted in Figure 8.5. The sample was initially saturated by Sea-water, drained, aged and then spontaneously imbibed by Diluted-sea-water. At the beginning of the forced imbibition, the water saturation of sample In17 was 42.4 %. The production plot of the centrifuge experiment shows that the applied nine spine velocity steps between 1000 to 7500 RPM caused

8.5 SIMULATION RESULTS

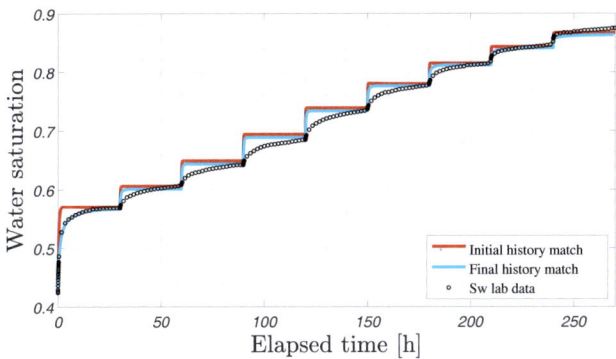

Figure 8.4: Comparison of initial and final centrifuge history match - The specification of the analytically obtained capillary inlet pressure matched in the endpoint water saturations. The relative permeability adjustment either delayed or accelerated oil production.

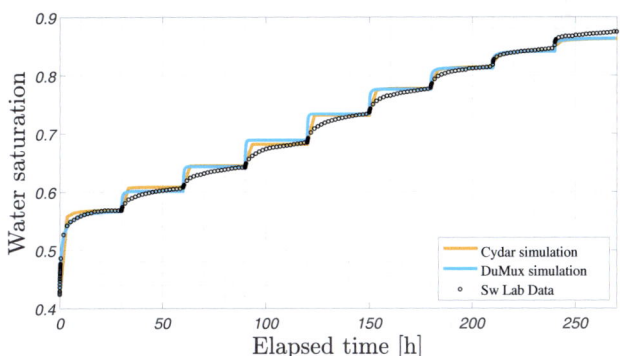

Figure 8.5: Centrifuge history match In17 - Cydar and DuMu^x both resulted in an accurate match of the experimentally obtained oil production.

the step-wise water saturation increase. The final average water saturation of 87.4% was obtained at a centrifuge spin of $7500\,RPM$.

In addition to the water displacement development illustration of Figure 8.6, a qualitative analysis of the In17 DuMu^x history matching simulation is plotted in Figure 8.7 to Figure 8.9. In line with Figure 8.2 and Figure 8.6, the inlet face is located on the right-hand side.

8.5 SIMULATION RESULTS

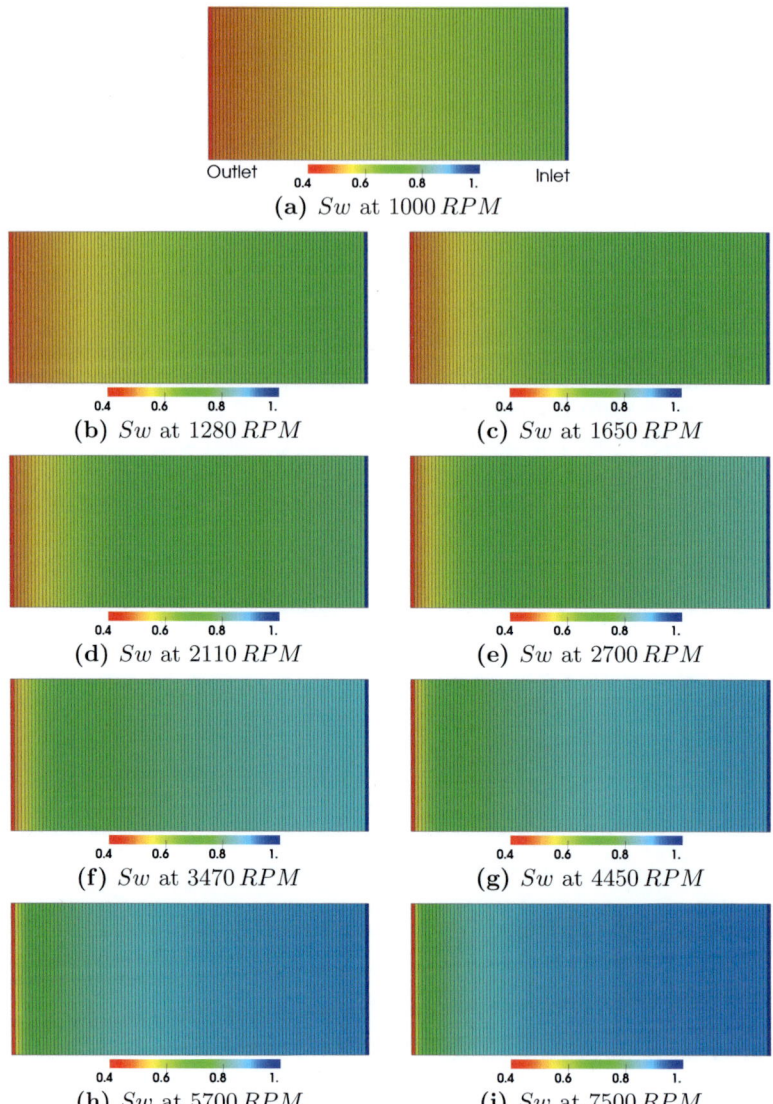

Figure 8.6: DuMu$^\text{x}$ centrifuge simulation In17 - Sw of In17 at the end of the corresponding centrifuge spin. The inlet face is located on the right-hand side while the outlet face is located on the left-hand side. The simulation uses a Cartesian 100 x 1 grid.

8.5 SIMULATION RESULTS

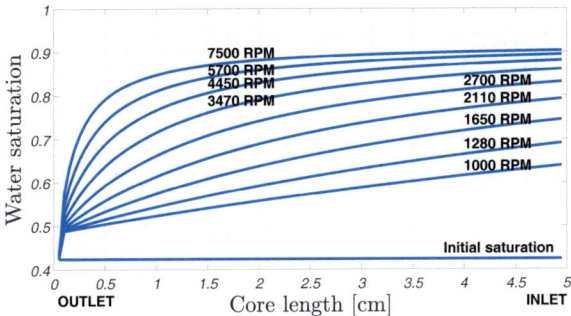

Figure 8.7: Sw profile In17 - The Sw illustrates the heterogeneous saturation distribution of the centrifuge method. The discrepancy between the inlet and average water saturation decreases with increasing centrifuge spin.

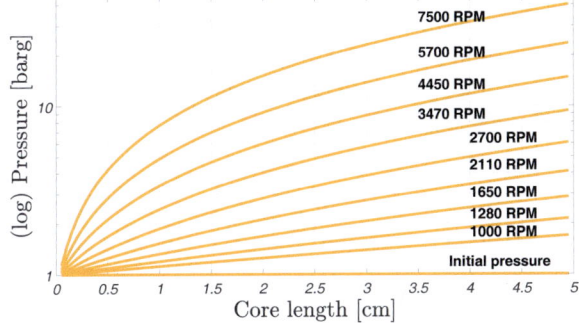

Figure 8.8: Pressure profile In17 - The pressure profile illustrates the pressure drop along the core as a function of applied centrifuge spin.

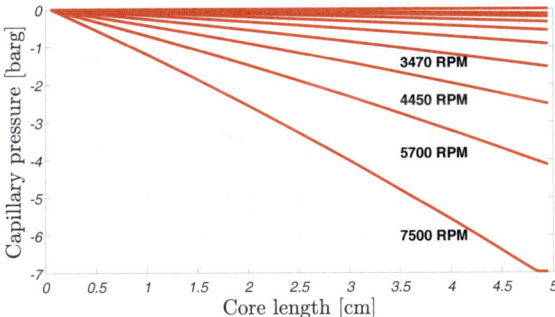

Figure 8.9: Capillary pressure profile In17 - In line with the water saturation, the arising capillary decreases towards the core outlet.

131

8.5 SIMULATION RESULTS

Figure 8.7 reflects the typical saturation profile development during the centrifuge experiment. While the highest water saturation is obtained at the inlet face, the water saturation decreases towards the outlet face. It is an illustrative example of the necessity of conducting an average to inlet water saturation correction. The saturation discrepancy between the numerically determined inlet and experimentally obtained average saturation amounts 9.8 % at a centrifuge spin of 2700 RPM. Once the centrifuge spin is increased, the difference between inlet and average saturation decreases towards 3.7 % at a final RPM of 7500.

Besides illustrating the saturation heterogeneity, Figure 8.7 shows the compliance of the Dirichlet boundary condition criterium. While the inlet boundary face is not plotted (constant Sw of 100 %), the outlet water saturation remains at the imposed water saturation of 42.4 %. In line with the Hassler & Brunner conditions, a water breakthrough is not observed.

The water phase pressure development is plotted in Figure 8.8, in which the y-axis is plotted on a logarithmic scale to improve the pressure representation. Despite the first impression of the illustration, the pressure drop along the core follows a linear correlation. In line with the water saturation, Figure 8.8 reflects the compliance of the selected Dirichlet boundary conditions. While Equation 8.23 precisely calculates the arsing inlet pressure, the outlet pressure remains at the defined pressure of 1 bar. The linear pressure behavior along the core furthermore implies the correct implementation of the centrifugal force.

A cross-section of the arising capillary pressure is plotted in Figure 8.9. While the highest capillary pressure is obtained at the inlet face (peak water saturation), the outlet capillary pressure remains at zero. Moreover, Figure 8.9 implies, that the Cydar inlet boundary formulation is in line with the imposed DuMu$^\text{x}$ Dirichlet conditions. While Cydar directly calculates the arising capillary pressure (cf. Equation 8.39), the DuMu$^\text{x}$ model calculates the water phase pressure (cf. Equation 8.23). Since capillary pressure is defined as differential phase pressure between the oil and water phase, both

8.5 SIMULATION RESULTS

boundary conditions are yielding into the identical capillary inlet pressure P_{ci}. Therefore, Equation 8.40 is valid for the Cydar and DuMu$^\text{x}$ inlet face boundary

$$P_{ci} = \frac{1}{2} \cdot \Delta\rho \cdot \omega^2 \cdot \left(r_{outlet}^2 - r_{inlet}^2\right). \tag{8.40}$$

8.5.3 Numerical capillary pressure

A comparison of the analytically and numerically obtained capillary pressure of sample In17 is plotted in Figure 8.10. The analytical solution of In17 corresponds to the Forbes corrected average to inlet water saturation correction (Equation 6.16), while the numerical capillary pressure is derived by the DuMu$^\text{x}$ history matching. The numerical reproduction of the centrifuge method confirms the validity of the experimentally obtained data. Furthermore, the history matching and the numerically obtained capillary pressure confirm the residual oil saturation at a centrifuge spin of $7500\,RPM$.

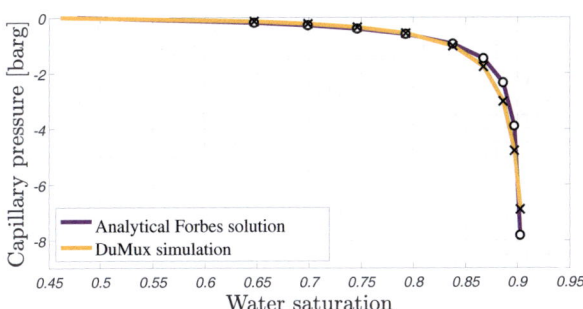

Figure 8.10: Analytical and numerical capillary pressure comparison - The comparison shows a good agreement between analytically and numerically obtained capillary inlet pressure.

8.6 Summary & conclusions

The chapter developed a numerical centrifuge model to history match the experimental data. Initially, the principle of the fully-implicit numerical DuMu$^\text{x}$ scheme was introduced. Besides the spatial and temporal discretization of the material balance, Section 8.1.1 to 8.1.3 introduced Newton's methods. The centrifuge model formulation included the description of the boundary conditions, fluid properties, hydraulic model, flow equation and centrifugal force implementation. The developed model was successfully validated against the commercial Cydar software.

- The implemented Dirichlet boundary conditions are suitable to represent the imbibition experiments. The imposed centrifugal acceleration is calculated for each grid node.

- The default DuMu$^\text{x}$ capillary pressure model was exchanged by an imbibition case adapted Modified-P_c model. The implemented Modified-P_c helped to obtain a satisfying history match. Moreover, the implemented Modified-kr model significantly improves the control of the input relative permeability curves, which allows improved matching of the saturation curvature.

- The numerical simulation results validate the analytically obtained capillary pressure. The obtained history matching confirmed the experimentally obtained residual oil saturation.

- The impact of the numerical relative permeability on the history matching was limited on the shape of the saturation curvature. While some studies are using the centrifuge method to derive the effluent relative permeability curves numerically, this study conducted unsteady state corefloodings experiments to derive relative permeability data.

9

Numerical coreflooding simulation

A two-phase numerical centrifuge model was presented in the previous chapter. Since the centrifuge experiments did not include an exchange of the imbibing brine, the fluid density and viscosity properties were defined by constant values. Compared to the application of a compositional fluid system, the simple structure of an immiscible two-phase system provides higher numerical stability. In the case of the centrifuge simulation, numerical stability is a crucial requirement to guarantee the compliance of the imposed Dirichlet boundary conditions.

As a result of the injection brine exchange during coreflooding In2b and In4b, the viscosity change significantly affected the emerging differential pressure. The two-phase system was therefore extended to a two-phase-three-component fluid system to consider the water viscosity and density reduction during the corefloodings. The oil phase is thereby consisting of an oil component that defines the (oil) phase properties by a constant viscosity and density value. Besides the water component, the water phase furthermore contains an aqueous salt tracer component to allow a dynamic water phase property definition. Thereby, the experimentally measured Formation-water, Sea-water and Diluted-sea-water viscosity and density values

9.1 DuMu^x COREFLOODING SIMULATION

are implemented as a function of the salt tracer component. As a result of the compositional fluid model introduction, the mathematical coreflooding formulation changes from mass concentration to mole fraction.

9.1 DuMu^x coreflooding simulation

In line with the centrifuge simulation, the coreflooding simulation is based on a fully-implicit numerical scheme. The cell-centered finite volume discretization is used as a spatial discretization scheme while the time discretization is based on the Backward Euler method. The system is assumed to be isothermal and a 2D 100 x 1 Cartesian grid represents the cores. Furthermore, the dynamic injection brine compositions were considered by the introduction of a salt tracer component. The tertiary oil recovery of coreflooding In4b was simulated by a linear interpolation between two relative permeability sets [24].

9.1.1 Boundary conditions

The principle of the coreflooding model is sketched in Figure 9.1. While the centrifuge model used a combination of Neumann and Dirichlet boundary conditions, the proposed numerical coreflooding model is limited to the application of Neumann boundaries. Each boundary face is consequently defined by a mathematical formulation of a source term (inlet face) or sink term (outlet face).

Figure 9.1 summarizes the mathematical inlet and outlet face formulation. In the case of the inlet face, the Neumann boundary formulation is defined by the experimentally applied injection rates. However, the applied volumetric rates of $0.05, 0.2, 1, 3$ and 5 cm^3/min required the adaption to the numerical 2D model. The required unit of $[kmol/(s \cdot m^3)]$ was obtained by inserting the volumetric brine injection rates into Equation 9.1

$$q_w^w = q_{lab} \cdot \frac{10^{-6} m^3}{60 s} \cdot \frac{\rho_w}{M_w A_{inlet}}, \qquad (9.1)$$

9.1 DUMU^X COREFLOODING SIMULATION

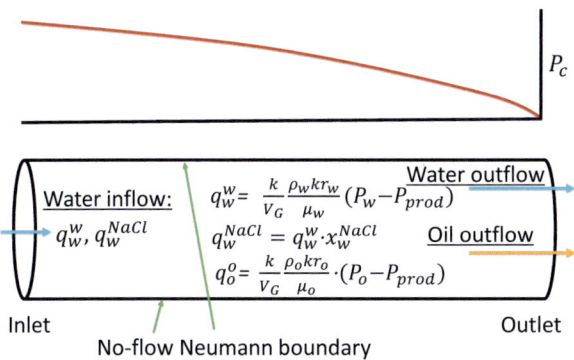

Figure 9.1: Numerical coreflooding model - The 2D injection rates define the water inflow, where the salt mole fraction is considered by q_w^{NaCl}. The outlet face mass extraction is calculated based on the phase mobilities and the experimentally set back-pressure.

where q_w^w is the 2D injection rate of the aqueous water component in $[kmol/(s \cdot m^3)]$, q_{lab} is the volumetric laboratory injection rate in $[cm^3/min]$, ρ_w is the mass density of the water phase in $[kg/m^3]$, M_w is the molar mass of the water phase in $[kg/kmol]$ and A_{inlet} is inlet face area in $[m]$.

Depending on the injection brine, the water phase had a salinity of $183.4\,g/l$ (Formation-water), $43.8\,g/l$ (Sea-water) or $0.438\,g/l$ (Diluted-sea-water). The salinity of the injection brine was considered by the admixture of a salt tracer (NaCl) component into the water phase. Since the salt component has identical properties as the water component, the admixture of the salt tracer component into the water phase did not change the molar balance. Consequently, the salt tracer component impact is limited to the interpolation of the brine density and viscosity. The formulation of the inlet face is completed by considering the injection brine salinity

$$q_w^{NaCl} = q_w^w \cdot x_w^{NaCl}, \qquad (9.2)$$

where q_w^{NaCl} is the 2D injection rate of the aqueous salt component in $[kmol/(s \cdot m^3)]$. The salt tracer mole fraction x_w^{NaCl} is defined by

9.1 DUMU^X COREFLOODING SIMULATION

the injection brine: 0.1834 (Formation-water), 0.0438 (Sea-water) and 0.000438 (Diluted-sea-water).

While the amount of injected mass at the inlet face is defined by the applied injection rates, the mathematical formulation of the production boundary is significantly more complex. The difficulty of realizing a reliable coreflooding simulation is represented by defining an outlet Neumann boundary condition, which authentically calculates the two-phase-three-component outflow production with respect to the experimentally set back-pressure. During the development of the coreflooding model, several boundary condition formulation approaches were tested. A simplification and adaption of Peaceman's well model revealed to be representative of the mass outflow. Equation 9.3 denotes Peaceman's well model in its most generalized form [24]

$$Q = \frac{2\rho k h_z}{\mu \left(\ln \left(\frac{r_e}{r_w} \right) + s \right)} \cdot (P - P_{bh}), \quad (9.3)$$

where Q is the injection or production rate in $[kg/s]$, ρ is the fluid density in $[kg/m^3]$, k is the permeability of the grid cell $[m^2]$, h_z is height of the grid cell in $[m]$, μ is the fluid viscosity in $[Pa \cdot s]$, r_e is the effective wellbore radius in $[m]$, r_w is the geometrical well radius in $[m]$ and s is the wellbore skin factor. Furthermore, P is the phase pressure in $[Pa]$ and P_{bh} denotes the bottom hole pressure in $[Pa]$.

In the case of a coreflooding experiment, all well related terms can be eliminated. The implemented outflow boundary condition for the oil and water phase was formulated as follows

$$q_w^w = \frac{k}{V_G} \cdot \frac{\hat{\rho}_w k r_w}{\mu_w} \cdot (P_w - P_{Prod}), \quad (9.4)$$

and

$$q_o^o = \frac{k}{V_G} \cdot \frac{\hat{\rho}_o k r_o}{\mu_o} \cdot (P_o - P_{Prod}), \quad (9.5)$$

9.1 DUMU^X COREFLOODING SIMULATION

where q_w^w is the production rate of the water component inside the water phase in $[kmol/m^2 s]$, q_o^o is the production rate of the oil component inside the oil phase in $[kmol/m^2 s]$, k is the absolute permeability in $[m^2]$, V_G is the grid volume in $[m^2]$, $\hat{\rho}_i$ is the fluid density in $[kmol/m^3]$, μ_i is the fluid viscosity in $[Pa \cdot s]$, kr_i the relative permeability, P_{Prod} is the production or back-pressure in $[Pa]$ and P_i is the grid phase pressure in $[Pa]$. The subscripts o and w denote the oil and water phase.

The quantity of the sink term is thereby mainly affected by the occurring phase mobilities and the experimentally set production pressure (back-pressure). Moreover, the water phase salt tracer component production is in balance with the water phase extraction

$$q_w^{NaCl} = \underbrace{\frac{k}{V_G} \cdot \frac{\rho_w kr_w}{\mu_w} \cdot (P_w - P_{Prod})}_{q_w^w} \cdot x_w^{NaCl}. \tag{9.6}$$

At a first glance, Equation 9.6 looks similar to the inlet face salt tracer component formulation (Equation 9.2). However, while the salt tracer component mole fraction at the injection face is statically defined by the injected brine composition (0.1834 for Formation-water, 0.0438 for Sea-water and 0.000438 for Diluted-sea-water), the salt tracer mole fraction at the outlet face is the numerically calculated salt mole fraction x_w^{NaCl}.

The source code implementation of the imposed Neumann inlet and outlet boundary is displayed in Appendix B.5 and Appendix B.4. The boundary formulation is completed by defining no-flow Neumann boundaries along the core.

9.1.2 Fluid properties

The density and viscosity of the Formation-water, Sea-water and Diluted-sea-water were experimentally determined at 70°C. The introduction of the salt tracer component allowed the density and viscosity definition as a function of salinity. After specifying the salinity of the injection brine inside the input file, the two-phase-three-

9.1 DUMU^X COREFLOODING SIMULATION

component system calculates the fluid properties as determined during the experimental study. In case the injection brine is exchanged, a quadratic polynomial function automatically correlates the experimentally measured fluid properties to the corresponding salinity (cf. Figure 9.2)

$$\rho_w(x_w^{NaCl}) = -1075.5 \cdot \left(x_w^{NaCl}\right)^2 + 896.3 \cdot \left(x_w^{NaCl}\right) + 975.3, \quad (9.7)$$

and

$$\mu_w(x_w^{NaCl}) = 0.819 \cdot \left(x_w^{NaCl}\right)^2 + 0.959 \cdot \left(x_w^{NaCl}\right) + 0.420, \quad (9.8)$$

where ρ_w is the brine density as function of the salt mole fraction x_w^{NaCl} and μ_w is the brine viscosity as a function of the salt mole fraction x_w^{NaCl}. The corresponding density and viscosity functions are plotted in Figure 9.2.

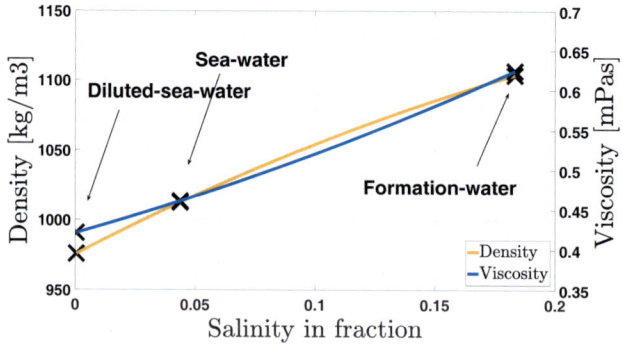

Figure 9.2: Viscosity and density at 70°C - The polynomial functions correlate the experimentally measured fluid properties to the brine salinity.

The implementation of a multi-phase and multi-component system usually requires a careful and time-consuming phase composition

9.1 DUMU$^\text{X}$ COREFLOODING SIMULATION

computation. Since the default DuMu$^\text{x}$ models are typically processing gas-liquid systems, DuMu$^\text{x}$ provides a complex miscible multi-phase composition solver. Based on the component-specific properties such as the Fugacity coefficient (gaseous phase) or Henry constant (liquid phase), the solver considers the phase miscibility and calculates the phase concentration and component fractions at each time step. Since the oil and water phases are assumed to be immiscible, the computation-intensive compositional solver is replaced by a manual composition definition.

As a result of the oil-water immiscibility, the oil phase is only consisting of the oil component

$$x_o^o = 1, \qquad (9.9)$$

where x_o^o represents the oil component mole fraction within the oil phase. Furthermore, the oil phase is restricted from dissolving into the water phase. Consequently, the remaining two possible water phase components are the water x_w^w and salt mole fraction x_w^{NaCl}. The water composition is solved by taking advantage of the fundamental model formulation. To solve the system of equations, DuMu$^\text{x}$ requires the definition of three primary variables. Besides the water pressure P_w and oil saturation So as the first two primary variables, the salt mole fraction x_w^{NaCl} represents the third primary variable. This allows a simple and efficient water phase composition formulation

$$\underbrace{\left(1 - x_w^{NaCl}\right)}_{x_w^w} + x_w^{NaCl} = 1, \qquad (9.10)$$

where x_w^w denotes the aqueous water component in mole fraction and x_w^{NaCl} represents the salt tracer mole fraction within the water phase.

9.1.3 Hydraulic properties

The capillary pressure P_c in $[Pa]$ denotes the pressure difference between the oil P_o and water phase P_w

9.1 DUMU$^\text{X}$ COREFLOODING SIMULATION

$$P_c(S_w) = P_o - P_w. \quad (9.11)$$

In accordance with the earlier developed centrifuge model, a hyperbolic function replaces the traditional Brooks-Corey capillary pressure model [30]. It is suitable to reflect the shape of a typical forced imbibition process, in which the water saturation increase causes an increase of the capillary pressure

$$P_c = \frac{B + C \cdot Sw_e}{1 + D \cdot Sw_e}. \quad (9.12)$$

The benefits of the Modified-kr model are discussed in Chapter 8. The replacement of the Corey exponents by the parameters ϵ, H and G allows a larger control of the relative permeability curves [27]. For completeness, the Modified-Corey model is listed in Equation 9.13 to 9.14

$$kr_w(S_w) = kr_w^{max} \cdot \left(\frac{e}{2\epsilon} \cdot Sw_e^{2\epsilon} + \frac{f}{\epsilon} \cdot Swe^e \cdot Sw_e^H\right), \quad (9.13)$$

and

$$kr_o(Sw) = kr_o^{max} \cdot \left(\frac{e}{2\epsilon} \cdot (1 - Sw_e)^{2\epsilon} + \frac{f}{\epsilon} \cdot (1 - Sw_e)^e \cdot (1 - Sw_e)^H\right), \quad (9.14)$$

In the case of coreflooding In4b and In9b, an interpolation between an oil-wet and a less oil-wet relativity permeability data set ensured the simulation of the additional oil recovery. The interpolation is implemented as a function of time, which is discussed in detail in Section 9.3

$$\epsilon_i^m(t), \ H_i^m(t), \ G_i^m(t), \quad m = 1, 2, \quad i = o, w, \quad (9.15)$$

9.1 DUMU$^\text{X}$ COREFLOODING SIMULATION

where ϵ_i^m, H_i^m and G_i^m are the input parameters of the two modified relative permeability model sets. The subscript i denotes the oil and water phase and the superscript m denotes an oil-wet (1) and a less oil-wet (2) relativity permeability data set.

9.1.4 Flow equation

In line with the centrifuge model, the advective flux v_i is calculated based on Darcy's law

$$v_i = -\frac{k k r_i}{\mu_i} \cdot (\nabla P_i - \hat{\rho}_i g), \quad i = o, w, \tag{9.16}$$

where k is the absolute permeability in $[m^2]$, kr_i is the relative permeability, P_i is the molar phase pressure in $[Pa]$, $\hat{\rho}_i$ is the phase density in $[kmol/m^3]$. While g denotes the centrifugal acceleration in the centrifuge model, the coreflooding model includes the standard gravitational acceleration of -9.81 m/s^2.

Besides the advective flux, the two-phase-three-component system required the consideration of the internal phase component diffusion. Since the miscibility between the oil and water phase was set to zero, the molecular diffusion is eventually limited to the salt tracer component inside the water phase. The salt tracer diffusion within the water phase is calculated by Fick's law (9.17)

$$J_w^{NaCl} = -(\rho_w \phi S_w \tau \tilde{D}_{\text{diff},w}^{NaCl}) \cdot \nabla x_w^{NaCl}, \tag{9.17}$$

where J_w^{NaCl} is the diffusive flux in $[kmol/(s \cdot m^3)]$, τ is the tortuosity of the porous medium and $\tilde{D}_{\text{diff},w}^{NaCl}$ is the salt binary diffusion coefficient inside the water phase $[m^2/s]$.

Ghaffari et al. [42] concluded that the ion self-diffusion coefficient is impacted by temperature and ion concentration. In general, the diffusion coefficient increases with increasing temperature and decreasing ion concentration. In regard to the isothermal coreflooding

9.1 DUMU^X COREFLOODING SIMULATION

experiment, the diffusion coefficient was therefore implemented as a function of the salt mole fraction. In accordance with the density and viscosity measurements, a quadratic polynomial function was used to correlate the diffusion values to the salinity. The experimentally determined diffusion values were defined as suggest by Ghaffari et al. [42]

$$\tilde{D}_{\text{diff},w}^{NaCl}(x_w^{NaCl}) = \left(8.0 \cdot \left(x_w^{NaCl}\right)^2 - 11.4 \cdot \left(x_w^{NaCl}\right) + 3.2\right) \cdot 10^{-9}, \quad (9.18)$$

where $\tilde{D}_{\text{diff},w}^{NaCl}$ is the binary diffusion coefficient of the salt tracer inside the water phase as a function of salinity x_w^{NaCl}.

9.1.5 Molar balance

The introduction of the salt tracer component required the model extension to a two-phase-three-component system. In contrast to the two-phase centrifuge model, the two-phase-three-component coreflooding model processes mole fraction instead of mass fractions. The molar balance can be summarized as follows

$$\phi \frac{\partial \left(\hat{\rho}_o x_o^k S_o + \hat{\rho}_w x_w^k S_w\right)}{\partial t} + \nabla \cdot \left(\hat{\rho}_w x_w^k v_w + J_w^k + \hat{\rho}_o x_o^k v_o\right) = q_i^k, \quad (9.19)$$

$$i = o, w \quad k = o, w, NaCl,$$

where ϕ is the porosity, $\hat{\rho}_i$ is the molar density in $[kmol/m^3]$, x_i^k is the mole fraction of the components inside the respective fluid phase, S_i is the phase saturation, v_i is the advective volumetric velocity in $[m/s]$ and J_i is the diffusive flux in $[kmol/(s \cdot m^2)]$ and q_i^k is the component source/sink term within the respective fluid phase in $[kmol/(s \cdot m^2)]$. The subscript i denotes the oil and water phase while the superscript k refers to the oil, water and salt tracer component.

The system of equations is closed by the sum of saturations and mole fractions

$$S_o + S_w = 1, \quad (9.20)$$

9.2 CYDAR COREFLOODING SIMULATION

and

$$x_o^k + x_w^k = 1. \quad (9.21)$$

The equations are discretized by using a cell-centered finite volume scheme as spatial discretization and the implicit Euler method as temporal discretization. By combing the constitutive relations for the capillary pressure (Equation 9.12), relative permeability (Equation 9.13 and Equation 9.14) and using the closure Equation 9.20 and Equation 9.21, the number of unknowns is reduced to three (Sw, Po and x_w^{NaCl}).

9.1.6 Time manager

In regard to the dynamic Neumann boundary conditions, the simulation is divided into a sequence of time episodes. After the problem initialization, the time manager controls the time step size of the simulation. In order to ensure the correct temporal representation of each injection rate, the change of the injection rate is reflected by the initialization of a new time episode.

9.2 Cydar coreflooding simulation

9.2.1 Cydar input data

The history matches of the Cydar and DuMu^x coreflooding simulations are based on the identical input data. Besides the fluid densities and viscosities at 70°C, Cydar furthermore required the definition of the core dimensions, porosity, permeability and initial water saturation. The applied brine injection rates were specified in cm^3/min and automatically adapted to the 2D simulation grid. Relative permeability and capillary pressure input data were required to start the numerical simulation.

In contrast to the proposed DuMu^x coreflooding model, the used Cydar version (8.1.1.0) did not support a viscosity/salinity correlation. Consequently, the Cydar history matches do not reflect the ex-

9.3 SIMULATION RESULTS

perimentally obtained differential pressure as the injection brine was exchanged. The Cydar coreflooding simulation is based on a fully-implicit incompressible numerical scheme. In line with the DuMu^x simulation, a 2D 100 x 1 cell grid was selected to represent the core.

9.2.2 Cydar boundary conditions

The work of Lenormand [60] described the principle of an incompressible unsteady state brine imbibition Cydar coreflooding model as following. After starting the coreflooding simulation, the outlet boundary condition initially only allows the production of oil. Thereby, the amount of produced oil is identical to the amount of injected brine. After water breakthrough at the outlet face, the capillary outlet pressure remains at zero and the water and oil phase are co-produced.

9.3 Simulation results

9.3.1 Absolute permeability matching

Before simulating the complex two-phase coreflooding experiments, the numerical coreflooding model was validated against a more simple single-phase reference case. The conducted absolute brine permeability measurements represented thereby an ideal opportunity to evaluate the validity of the numerical model.

Assuming a correctly implemented numerical model, the experimentally and numerically obtained differential pressure of the absolute brine measurements should be in close agreement. To ensure the differential pressure comparison, the core properties, fluid properties and permeability values were specified as measured during the experimental work.

The experimentally and numerically obtained differential pressure values of the absolute brine permeability measurements of sample In2b are listed in Table 9.1. The discrepancy between the numerical and experimental determined differential pressure ranges between 1.2 to 2.3 %. It is an overall fairly close agreement, which indicates

9.3 SIMULATION RESULTS

that the developed coreflooding model is valid for a single-phase flow case. In particular, the volumetric 3D to 2D rate conversion (Equation 9.1) are correctly implemented. Furthermore, the imposed outlet boundary conditions (Equation 9.4 to Equation 9.6) resulted into an adequate mass extraction. The outlet pressure remained at the set back-pressure of 6.4 bar (100 psi).

		Flowrate $[cm^3/min]$	Meas. ΔP $[barg]$	Num. ΔP $[barg]$	Deviation [%]
Sample	In2b	1	0.714	0.723	1.2
Length [mm]	37.95	1.5	1.064	1.085	2.0
Diameter [mm]	48.68	2	1.427	1.447	1.4
Viscosity [mPas]	1.45	2.5	1.767	1.808	2.3

Table 9.1: Single phase coreflooding model validation - The difference between the experimental and numerical differential pressure values ranges between 1.2 and 2.3 %. The good agreement indicates the validity of the numerical model.

9.3.2 History matching methodology

In theory, the injection at field rate equivalent leads to the final remaining oil saturation. The concept of critical capillary number assumes that a subsequent remaining oil saturation reduction only occurs in case the critical capillary number is exceeded. The exceeding of the critical number might be caused by the significant increase of the flow velocity (1), an increase of the displacing fluid viscosity (2), or the appearance of ultra-low IFT values (3) [1, 58] (cf. Equation 7.2).

In a dissent to the concept of capillary trapping, it has been shown that especially strongly oil-wet/water-wet systems are significantly impacted by capillary end effects. In this case, rate bumping causes a reduction of the remaining oil saturation by overcoming the capillary end effect.

Mcphee et al. [72] suggested a remaining oil saturation decrease of up to five saturation units, in case the capillary end effect is mitigated due to rate bumping.

9.3 SIMULATION RESULTS

The work of Nasralla et al. [82] included several unsteady state reservoir limestone waterfloodings. Compared to the oil recovery at field rate equivalent injection, the work reported a recovery factor increase from approximately 30 % to approximately 60 % when the injection rates were bumped up.

A similar tendency was observed during the unsteady state corefloodings of Chapter 7, in which the quantity of the remaining oil saturation reduction showed a correlation to the injection brine composition. At an injection rate of $0.05\,cm^3/min$, the remaining oil saturation of Formation-water in secondary injection mode was 61 % and decreased to 37.4 % when the injection rate was bumped up to $5\,cm^3/min$. The secondary injection of Sea-water at $0.05\,cm^3/min$ caused a So_{rm} of 51.3 % and a So_{rm} of 35.5 % when increasing the injection rate to $5\,cm^3/min$. Moreover, the injection of Diluted-seawater at $0.05\,cm^3/min$ resulted into a So_{rm} of 46.2 % and So_{rm} of 30.6 % after increasing the injection rate to $5\,cm^3/min$.

Kamath et al. [58] compared the unsteady state coreflooding behavior of different carbonate core materials. In the case of limestones samples, a differential pressure gradient increase affected the microscopic trapping mechanism. The author concluded that in dissent to the critical capillary number concept, additional oil can be recovered without exceeding the critical capillary number. Consequently, the concept of the critical capillary number might not be applicable to limestones.

To obtain satisfying history matches of coreflooding In2b, In4b and In9b, the impact of rate bumping on oil recovery needs to be considered by the numerical model. Without adapting the boundary face conditions, the remaining oil saturation is otherwise immediately obtained at the first injection rate. This is, however, not representative of the capillary end effect and does not reflect the experimental data.

To represent the impact of rate bumping, the numerical core grid is extended by an additional cell that reflects the transition between the porous (core) and non-porous media (production line) dominated flow-environment. The additional grid cell is thereby characterized

9.3 SIMULATION RESULTS

by a large permeability and porosity value. Furthermore, the relative permeability follows a linear correlation and capillary pressure is set to zero [60, 63].

As a result of the introduction of an outlet cell, the dynamics of the numerical coreflooding simulation model substantially changes. Initially, the field rate equivalent injection recovers a certain amount of oil until the specified capillary pressure/relative permeability properties prevent further oil displacement. As the injection rate is increased, the increasing differential pressure overcomes the capillary pressure and hence mobilizes additional oil.

While the centrifuge history matching was predominantly impacted by the capillary input pressure, the history matching of the corefloodings required a careful relative permeability and capillary pressure adaption. At the beginning of the history matching, an intermediate-wet relative permeability (Corey of 2) was defined. The connate water saturation Sw_c and remaining oil saturation So_{rm} were thereby specified as experimentally measured. Moreover, the maximum (endpoint) oil relative permeability kr_o^{max} and maximum (endpoint) water relative permeability kr_w^{max} were defined in line with the coreflooding results (cf. Equation 7.3 and Equation 7.4). The initial capillary pressure curves were specified as measured during the centrifuge experiments.

The history matches were obtained by the step-wise modification of the relative permeability input parameters (ϵ_i, H_i and G_i, Equation 9.13 and Equation 9.14) as well as capillary pressure input parameters (B, C, D, Equation 9.12). The oil production at the first two injection rates of 0.05 and $0.2\,cm^3/min$ significantly impacted the relative permeability. While the small oil recovery of coreflooding In2b required the specification of a non-water wet relative permeability, the much higher oil recovery in the case of Diluted-sea-water injection (In9b) required the specification of a water-wet relative permeability (Figure 9.13). Furthermore, in order to match the obtained differential pressure, the initially defined (centrifuge method) capillary pressure data required a reduction in the range of 50 %. A detailed comparison of the numerically obtained rela-

9.3 SIMULATION RESULTS

tive permeability and capillary curves of coreflooding In2b, In4b and In9b are plotted in Figure 9.11 and 9.12.

The Cydar as well as DuMu^x simulation are using a 2D 100 x 1 Cartesian grid in which the permeability and porosity are homogeneously distributed. The plotted Cydar and DuMu^x simulation results of Figure 9.3, 9.4 and Figure 9.10 are based on the identical capillary pressure and relative permeability input data. Moreover, Figure 9.6 displays the selected DuMu^x 2D 100 x 1 Cartesian grid.

9.3.3 Formation-water in secondary mode

Figure 9.3 illustrates the water saturation and differential pressure history matches of coreflooding In2b. The results of the DuMu^x as well as Cydar simulation are plotted. The injection of Formation-water in secondary mode caused significant oil recovery at the injection rates of $0.05, 0.2$ and $1 \, cm^3/min$. The subsequent injection rates of 3 and $5 \, cm^3/min$ led into minor additional oil recovery, which corresponded to a final residual oil saturation of 37.4 %. During the secondary injection of Formation-water, both numerical approaches resulted in a reasonably close water saturation and differential pressure match.

Whereas the exchange of the Formation-water by Sea-water and Diluted-sea-water did not cause additional oil recovery, Figure 9.3 records a significant differential pressure reduction. The pressure drop is caused by the reduction of water viscosity. Compared to the viscosity of the initially injected Formation-water, the injection of Sea-water caused a viscosity reduction of approximately 26 %. Furthermore, the Diluted-sea-water is characterized by a 33 % smaller viscosity than the Formation-water. While Cydar did not support the consideration of the injection fluid viscosity change, the developed DuMu^x model defined the viscosity (and density) as a function of salinity. Equation 9.22 reflects the measured viscosity of Formation-water, Sea-water and Diluted-sea-water at 70°C (cf. Figure 9.2)

9.3 SIMULATION RESULTS

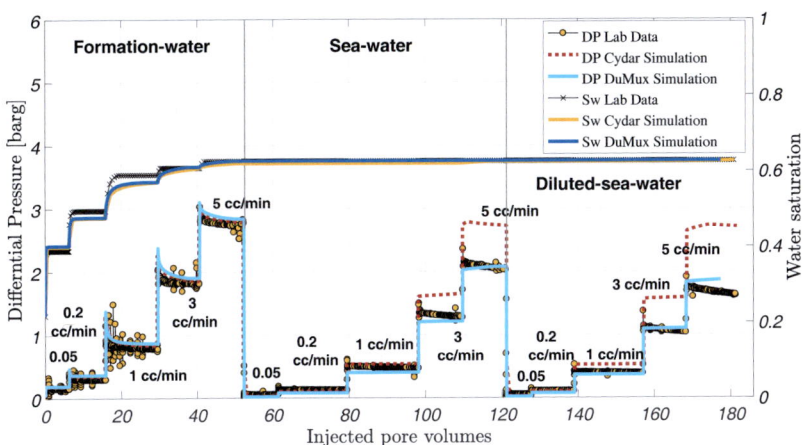

Figure 9.3: Numerical history match coreflooding In2b - Injection of Formation-water in secondary mode followed by the injection of Sea-water and Diluted-sea-water in tertiary injection mode.

$$\mu_w(x_w^{NaCl}) = 0.819 \cdot \left(x_w^{NaCl}\right)^2 + 0.959 \cdot \left(x_w^{NaCl}\right) + 0.420. \quad (9.22)$$

The analysis of Figure 9.3 indicates that the viscosity and salinity correlation is sufficient to obtain a suitable differential pressure match. While the Cydar software resulted in significant differential pressure overestimation, Equation 9.22 yielded into a precise pressure match during the secondary and tertiary brine injection.

9.3.4 Sea-water in secondary mode

The injection of Sea-water in secondary injection mode caused major oil production at the injection rates of 0.05 and 0.2 cm^3/min (Figure 9.4). A remaining oil saturation of 35.5 % was observed after the injection of Sea-water in secondary injection mode. In line with coreflooding In2b, the exchange of the injection brine caused a reduction of the emerging differential pressure. The viscosity of the tertiary injected Diluted-sea-water was approximately 9 % smaller

9.3 SIMULATION RESULTS

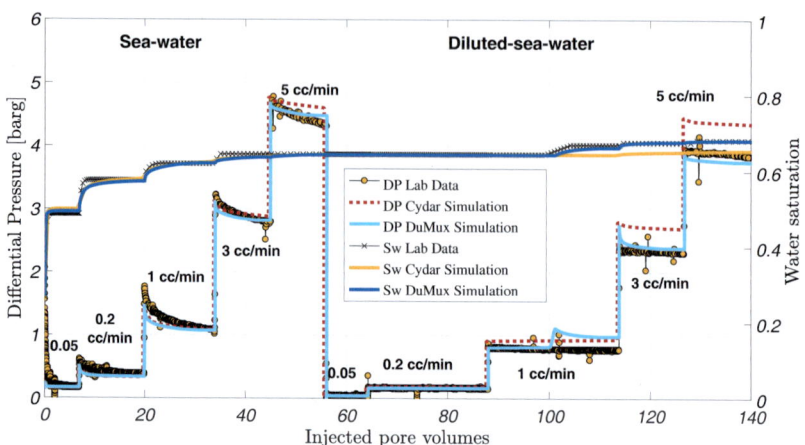

Figure 9.4: Numerical history match coreflooding In4b - Injection of Sea-water in secondary mode followed by the injection Diluted-sea-water.

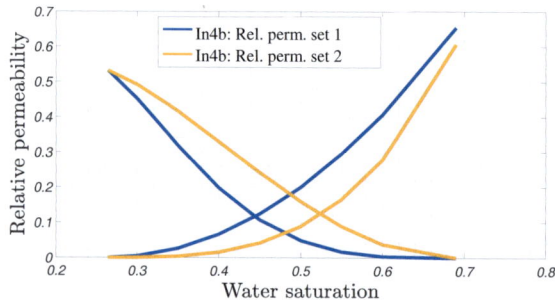

Figure 9.5: Relative permeability interpolation - Relative permeability comparison of coreflooding In4b. The interpolation between the two data sets allowed the simulation of the additional oil recovery of coreflooding In4b.

compared to the viscosity of the secondary injected Sea-water. The DuMu^x simulation resulted in a precise differential pressure match, while the static Cydar viscosity definition caused an overestimation of the differential pressure.

9.3 SIMULATION RESULTS

After the injection of roughly 100 pore volumes of injection fluid, Figure 9.4 shows a sudden oil recovery increase of 3.4%. The increase of the oil recovery might be caused by an experimental artifact or a low-salinity effect due to the injection of Diluted-sea-water in tertiary injection mode. Independent from the physical reason, the DuMu^x coreflooding model uses a linear relative permeability interpolation to simulate the additional recovery. The interpolation is based on the definition of two relative permeability sets: A less water-wet relative permeability data set which was used before, and a stronger water-wet relative permeability data set which was used after the injection of $100\,V_P$ of brine. While typical (low-salinity) relative permeability interpolations are typically implemented as a function of salinity, the injection history of coreflooding In4b did not justify this approach. Due to the injection of roughly $60\,V_P$ of Sea-water and $40\,V_P$ of Diluted-sea-water, the high-saline connate water was completely displaced from the homogeneous simulation model. To ensure the simulation of the additional oil recovery, the relative permeability interpolation was therefore imposed as a function of time. The transition between the two permeability sets was buffered by a time-depending linear correlation

$$\epsilon_i^m(t),\ H_i^m(t),\ G_i^m(t), \quad m = 1, 2, \quad i = o, w, \quad (9.23)$$

where ϵ_i^m, H_i^m and G_i^m are the input parameters of the two modified relative permeability model sets. The subscript i denotes the oil and water phase and superscript m denotes an oil-wet (1) and a less oil-wet (2) relativity permeability data set. The two interpolated relative permeability sets are displayed in Figure 9.5.

The history match of coreflooding In4b (Figure 9.4) shows that the imposed relative permeability interpolation is suitable to simulate the additional oil recovery. As a result of the increasing water saturation (after the injection of $100\,V_P$), the increasing capillary pressure causes an increase in differential pressure. The tertiary injection of Diluted-sea-water caused an overall oil recovery increase of $5.4\,\%$, which corresponded to a final residual oil saturation of $31.7\,\%$.

9.3 SIMULATION RESULTS

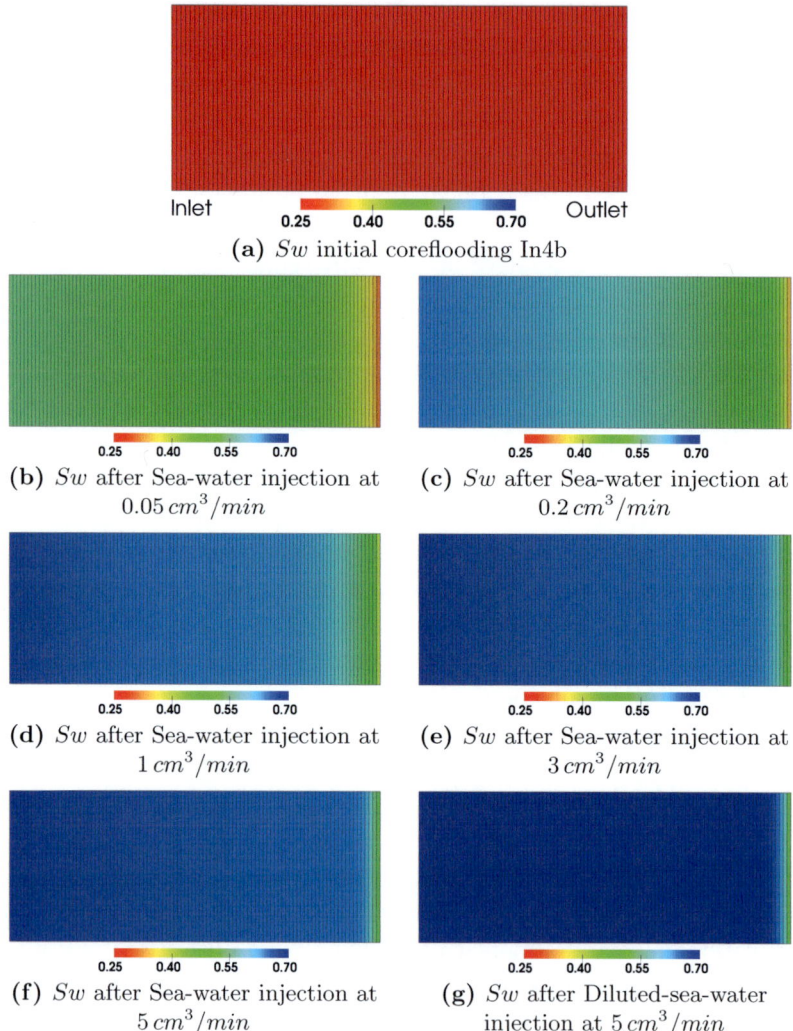

Figure 9.6: DuMu^x coreflooding simulation In4b - The Figures are showing the water saturation of coreflooding In4b at the ending of each injection rate. In line with the sketched coreflooding model (cf. Figure 9.1), the inlet boundary is located on the left-hand side while the outlet face is located on the right-hand side. The numerical simulation uses a 2D Cartesian 100 x 1 grid.

9.3 SIMULATION RESULTS

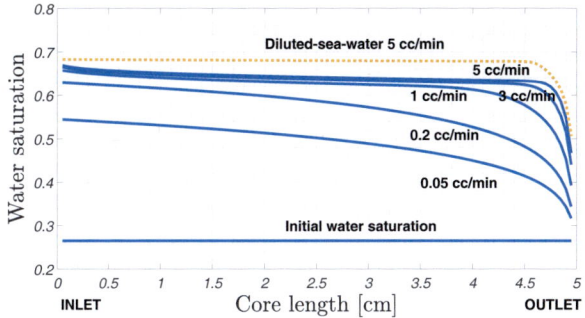

Figure 9.7: Sw profile In4b - The cross-section shows Sw at the end of each injection rate. The rate bumping causes a capillary end effect mitigation.

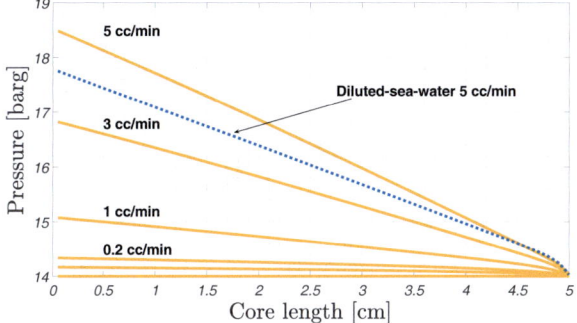

Figure 9.8: Pressure profile In4b - The inlet pressure linearly decreases towards the production face. Compared to Sea-water, the smaller viscosity of Diluted-sea-water results in a reduction of the phase pressure.

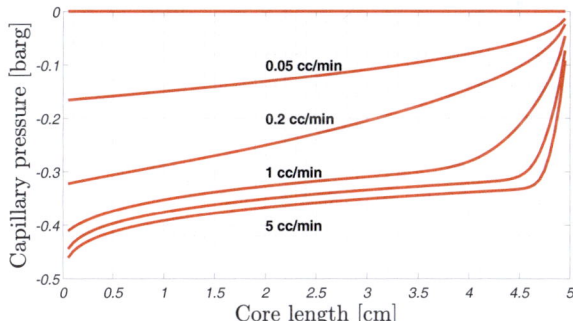

Figure 9.9: Capillary pressure profile In4b - The capillary pressure is a function of the water saturation.

155

9.3 SIMULATION RESULTS

Coreflooding In4b was selected to evaluate the imposed boundary conditions (Figure 9.7 to Figure 9.9). The water saturation development during the secondary injection of Sea-water is plotted in Figure 9.7, where the injection face is located on the left-hand side. Especially at the lower injection rates of $0.05, 0.2$ and $1\ cm^3/min$, the water saturation results into a heterogeneous distribution. While the peak water saturation occurs at the inlet face, the water saturation slowly decreases towards the outlet face. The characteristic saturation profile is caused by the definition of a zero capillary pressure outlet face condition. Besides representing the capillary end effect, the capillary pressure outlet face condition ($P_c = 0$) allows the consideration of the water saturation increase at each injection rate. As a result of the injection rate increase, the increasing differential pressure mobilizes additional oil by overcoming the capillary pressure/capillary end effect. Consequently, a new equilibrium between fluid saturations, relative permeability and capillary pressure establishes. In addition to the water cross-section of Figure 9.7, Figure 9.6 displays the water saturation development at 2D grid scale.

Besides the secondary Sea-water injection, Figure 9.7 illustrates the final water saturation after the injection of Diluted-sea-water at $5\ cm^3/min$. The stronger water-wet relative permeability is causing a fairly homogeneous water saturation increase and the attenuation of the capillary end effect.

The pressure development along the core is plotted in Figure 9.8. The highest pressure occurs at the injection face and linearly decreases towards the outlet. While the inlet pressure depends on the applied injection rates, the proposed outlet boundary condition (Equation 9.5 to Equation 9.6) remains at a constant production pressure of $14\ bar$. As a consequence of the viscosity reduction, the water phase pressure of the Diluted-sea-water phase decreases.

Figure 9.9 summarizes the capillary pressure development. The arising capillary pressure along the core is a function of the occurring water saturation. In line with the water saturation profile (Figure 9.7), the highest capillary pressure occurs at the inlet face and decreases towards the outlet. The outlet capillary pressure remains at zero.

9.3 SIMULATION RESULTS

9.3.5 Diluted-sea-water in secondary mode

The history match of the injection of Diluted-sea-water in secondary mode is plotted in Figure 9.10. Coreflooding In9b is characterized by a high oil recovery at the lowest injection rate of $0.05\,cm^3/min$. Since coreflooding In9b did not include the application of a tertiary injection brine, DuMu^x as well as Cydar resulted in accurate differential pressure match.

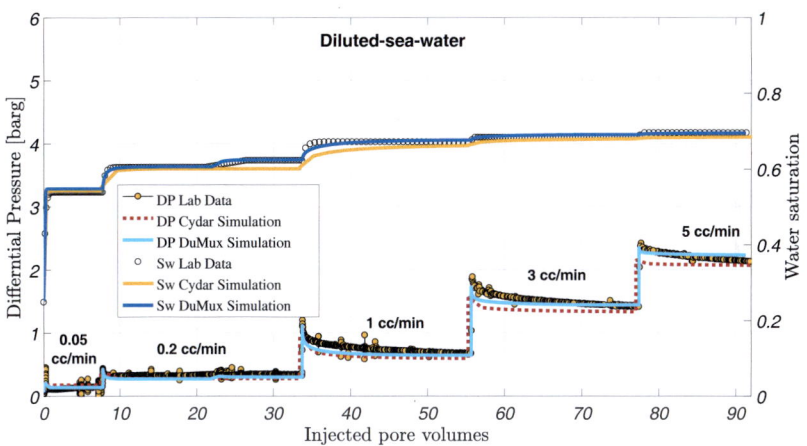

Figure 9.10: Numerical history match coreflooding In9b - Injection of Diluted-sea-water in secondary mode.

The injection of Diluted-sea-water at $0.2\,cm^3/min$ caused a water saturation increase after the injection of $22\,V_P$ of brine. In line with coreflooding In4b, this might be explained by an experimental artifact or by the additional oil mobilization due to low-salinity effects. Both cores were initially saturated by the high saline Formation-water. To simulate the water saturation increase, the DuMu^x simulation applies a relative permeability data set interpolation. The additional oil recovery is thereby simulated by defining a slightly stronger water-wet relative permeability curve after the injection of

9.3 SIMULATION RESULTS

22 V_P of Diluted-sea-water. The remaining oil saturation of coreflooding In9b was 30.6 %.

9.3.6 Relative permeability and capillary pressure

One of the primary purposes of conducting coreflooding experiments is the derivation of relative permeability data sets. The presented Modified-kr model includes three variables. While the quantities of the parameters H, G and ϵ were step-wise estimated during the history matching, the oil endpoint relative permeability kr_o^{max} and water endpoint relative permeability kr_w^{max} were calculated based on the experimental data. A detailed explanation is provided in Chapter 7.

The comparison of the production data of coreflooding In2b, In4b and In9b are showing significant oil recovery differences at an injection rate of 0.05 cm^3/min. Applying the field rate equivalent in secondary injection mode, Formation-water caused 21.9 % oil recovery, Sea-water caused 30.1 % oil recovery and Diluted-sea-water caused 38.5 % oil recovery (cf. Figure 9.13). Since the oil recovery at the lowest injection rates significantly impacted the shape of the numerically obtained relative permeabilities, the relative permeability curves of Figure 9.11 shows the strongest water-wetting behavior when Diluted-sea-water was injected in secondary mode.

This is line with the results of the spontaneous imbibition experiments, where Diluted-sea-water caused a 13.3 % higher average oil recovery than Sea-water and a 33.9 % higher average oil recovery than Formation-water. Furthermore, the centrifuge experiments showed the strongest water-wet behavior when Diluted-sea-water was used as imbibing fluid.

The capillary pressure input curves of coreflooding In2b, In4b and In9b are plotted in Figure 9.12. The capillary pressure data were initially specified as measured during the experimental part. However, to obtain a satisfying history match, the capillary input pressure data required a significant reduction in the range of 50 %. In accordance with the results of the centrifuge method, the numerically obtained coreflooding capillary pressure is showing the strongest

9.3 SIMULATION RESULTS

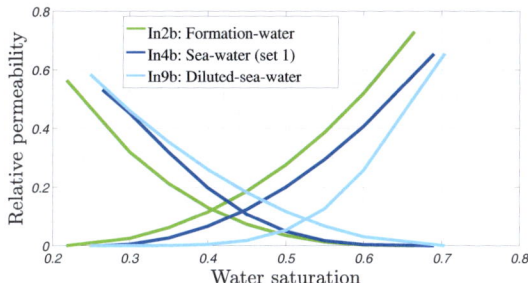

Figure 9.11: Relative permeability curves - The numerically obtained kr curves show the strongest water-wetting behavior for Diluted-sea-water, followed by Sea-water and Formation-water.

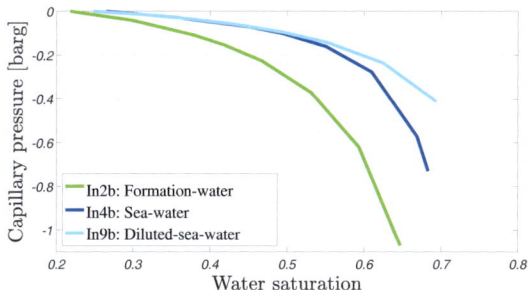

Figure 9.12: Capillary pressure curves - The numerically obtained Pc curves show the strongest water-wet behavior for Diluted-sea-water, followed by Sea-water and Formation-water.

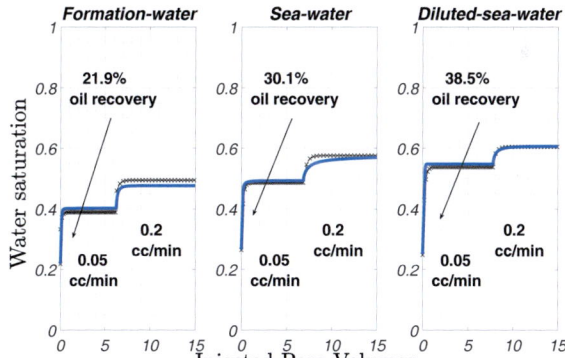

Figure 9.13: Field rate oil recovery - The field rate equivalent oil recovery significantly impacts the kr curves.

9.4 SUMMARY & CONCLUSIONS

water-wet conditions for Diluted-sea-water, followed by Sea-water and Formation-water (Figure 9.12).

Besides the relative permeability and capillary pressure curves, the analysis of the coreflooding experiments indicates an impact of injection brine on residual oil saturation. In secondary injection mode, Formation-water caused a remaining oil saturation of 37.4 %, Sea-water caused a remaining oil saturation of 35.5 % and Diluted-sea-water caused a remaining oil saturation of 30.6 %.

While the comparison of the coreflooding results in secondary application mode is consistent and in line with the spontaneous imbibition and centrifuge experiments, the analysis of the tertiary injection mode application is less consistent. Coreflooding In2b did not show additional oil recovery due to the injection of Sea-water and Diluted-sea-water in tertiary injection mode. Minor additional oil recovery of 5.4% was observed during coreflooding In4b when Diluted-sea-water was injected in tertiary injection mode. However, since the additional oil recovery was observed at the injection rates of 1.3 and 5 cm^3/min, the recovery increase might be predominantly caused by the mitigation of the capillary end effect.

The comparison of the remaining/residual oil saturation of the coreflooding and centrifuge experiments shows a large saturation discrepancy between the two methods. Based on the physics of the centrifuge and coreflooding experiments, it is expected that the endpoint saturations of the experiments are not identical. While the centrifuge is a gravity stable oil-water replacement process, the unsteady state coreflooding oil-water displacement is much more impacted by microscopic core heterogeneities, gravity, unfavorable mobility ratios and viscous fingering [72]. However, in regard to the magnitude of the discrepancy, further research such as the conduction of In situ saturation monitoring corefloodings is recommended.

9.4 Summary & conclusions

This chapter initially presented the mathematical formulation of the unsteady state coreflooding model. In regard to the injection

9.4 SUMMARY & CONCLUSIONS

brine exchange, a two-phase-three-component fluid model was implemented to reflect the exchange of the injection brines.

In the first place, the developed model was verified against absolute brine permeability measurements. The coreflooding history matches are showing a good agreement between the DuMu^x and Cydar simulation. The water saturation, water phase pressure and capillary pressure cross-section of coreflooding In4b indicate the compliance of the imposed boundary conditions. Furthermore, the history matches are confirming the physical plausibility of the experimentally obtained coreflooding data.

The results of the centrifuge method and USS corefloodings are overall in good agreement. The centrifuge method showed an increasing water-wetting tendency and a simultaneous reduction of the remaining oil saturation, as the salinity of the imbibing brines decreased. In regard to the coreflooding results, the injection of Diluted-sea-water at field rate equivalent caused by far the highest oil recovery. Additionally, when comparing the numerically obtained relative permeabilities, the injection of Diluted-sea-water in secondary is characterized by the strongest water-wet relative permeability curves.

- The presented numerical DuMu^x coreflooding model resulted in a reliable simulation of the conducted unsteady state corefloodings. The salt tracer component allowed the consideration of injection fluid viscosity changes. The proposed outlet boundary condition authentically calculated the mass outflow. Moreover, the implementation of the modified capillary pressure and modified relative permeability model helped to obtain satisfying history matches of coreflooding In2b, In4b and In9b.

- The good agreement between the Cydar and DuMu^x coreflooding simulation results demonstrate the validity of the proposed model. After defining an outlet capillary pressure of zero, the proposed boundary and fluid flow formulation result in a precise history matching.

9.4 SUMMARY & CONCLUSIONS

- In secondary injection mode, the numerically obtained relative permeability and capillary pressure curves showed the strongest water-wet behavior for the secondary injection of Diluted-sea-water, followed by Sea-water and Formation-water.

10

Summary & conclusions

The literature review initially summarized the current knowledge and understanding of low-salinity effects in limestones. It was pointed out that the literature focuses on two concepts to promote low-salinity effects. The modification of the overall potential determining ions and/or the lowering of the total ionic strength. The majority of the reported spontaneous imbibition experiments showed an improved oil recovery, in case the limestone samples were exposed to a highly diluted/desalinated water. While some coreflooding studies reported improved oil recovery in case of the injection of potential determining ions enriched brines, other studies confirmed the promising potential of highly diluted Sea-water to recover oil efficiently. The latest low-salinity research findings on limestone suggest Surface charge change as the driving mechanism of wettability alteration. Besides explaining the different low-salinity response of chalk, dolomite and limestone samples, the concept of Surface charge change emphasizes the potential of highly desalinated imbibing waters to increase oil recovery.

While the success of an oil recovery study is typically measured based on the experimental outcomes, the study preparation is a crucial prerequisite to obtain high-quality research data. This thesis therefore provides a detailed description of the fluid and core preparation. Moreover, the establishment of the initial water saturation and wettability properties are described. After using the cores in the experimental study, the reliability of the suggested core preparation and cleaning procedure was demonstrated by re-establishing similar sample conditions during a second preparation sequence.

In regard to the characteristic fracture-matrix systems of carbonate formations, spontaneous imbibition is an important oil recovery mechanism. To improve the understanding of the prerequisites of spontaneous imbibition in carbonates, the study tested the spontaneous imbibition behavior of eight different connate water and imbibing water brine combinations. While significant oil recovery was observed in case the imbibing water had a smaller salinity than the connate water, hardly any oil production was observed in case the imbibing water had a higher or identical salinity as the connate water. In the case of spontaneous oil recovery, the arising contact angles indicated intermediate to water-wet conditions. Furthermore, the conducted zeta potential measurements supported the concept of Surface charge change as the driving low-salinity mechanism in limestones. Compared to Formation-water and Sea-water, the Diluted-sea-water and limestone systems resulted in the significantly strongest negative surface charge.

For the first time in low-salinity limestone research (to the best of the authors knowledge), the centrifuge technique was comprehensively used to demonstrate the impact of salinity on wettability and residual oil saturation. In the first place, the Hassler & Brunner correction was derived to illustrate the necessity of conducting a reliable inlet to water saturation correction. It was furthermore shown that the combination of a hyperbolic regression and Forbes first solution leads into an efficient, unified and precise analytical correction of imbibition data. The acquired imbibition capillary pressure curves emphasized the promising potential of Diluted-sea-water to increase oil production. In the case of Formation-water as connate water, Diluted-sea-water as forced imbibition fluid showed the strongest water water-wet behavior and smallest residual oil saturations. Moreover, the combination of spontaneous and forced imbibition demonstrated that spontaneous imbibition experiments do not necessarily reveal evidence about Enhanced oil recovery. Although Diluted-sea-water as connate water and imbibing water did not result in any spontaneous oil recovery, the forced imbibition of Diluted-sea-water as connate water and imbibing water caused the smallest residual oil saturation.

In comparison to spontaneous imbibition and centrifuge method tests, unsteady state corefloodings are more representative of reservoir scale related displacement processes. After describing the fundamentals of the coreflooding technique, the concept of capillary end effect and rate bumping was introduced. At field rate equivalent injection, Diluted-sea-water caused a significantly higher oil recovery than Sea-water and especially Formation-water. Although the field rate equivalent injection typically interferes with capillary end effects, the results indicated the injection brine composition impact on wettability and relative permeability. After rate bumping, Diluted-sea-water showed a slightly higher oil recovery than Sea-water and Formation-water. The capability of Diluted-sea-water to enhance oil recovery in tertiary mode remained questionable, as minor additional oil recovery occurred in only one out of two tertiary injection applications.

The numerical part of this thesis includes the development of a $DuMu^x$ centrifuge and coreflooding model. After describing the principle of the fully-implicit numerical $DuMu^x$ scheme, a two-phase centrifuge model was formulated. The model description included the imposed Dirichlet boundaries and centrifugal force implementation. Furthermore, an imbibition process adapted Modified-P_c model as well as a Modified-kr model were presented. The imposed centrifuge model was successfully verified against the commercial Cydar software. While the numerical centrifuge simulation confirmed the analytical capillary pressure and residual oil saturations, the history matching questioned the validity of the numerically derived (centrifuge) relative permeability data. As a consequence, unsteady coreflooding experiments were conducted to obtain more reliable relative permeability data.

To simulate the tertiary injection brine exchange, the numerical coreflooding fluid model was extended to a two-phase-three-component system. In the first place, the developed coreflooding model was verified against absolute brine permeability measurements. Moreover, the coreflooding history matches showed a good agreement between the proposed model and the commercial reference simulation. While the inlet boundary formulation was defined by the experimentally applied injection rates, a mobility and back-pressure respecting production boundary model was proposed. After defining a constant outlet face capillary pressure of zero, the proposed DuMu^x coreflooding model resulted in a precise history matching. In secondary injection mode, the numerically obtained relative permeability showed the strongest water-wet behavior for Diluted-sea-water as injection brine.

- The spontaneous imbibition, centrifuge method and unsteady state coreflooding experiments demonstrated a correlation between salinity, wettability and oil recovery of limestone samples. Compared to Formation-water and Sea-water, Diluted-sea-water showed by far the most promising potential to efficiently recover oil. Using Formation-water as connate water, Diluted-sea-water resulted in 35.4 % spontaneous oil recovery, Sea-water in 22.1 % spontaneous oil recovery and Formation-water in 1.5 % spontaneous oil recovery. Moreover, the centrifuge method yielded into a residual oil saturation of 5.6 % for Diluted-sea-water, 10.7 % for Sea-water and 15.3 % for Formation-water. The unsteady state corefloodings confirmed the injection brine composition impact on oil recovery. While Diluted-sea-water caused the significantly highest oil recovery at secondary field rate injection, the remaining oil saturation after rate bumping was 30.6 % for Diluted-sea-water, 35.5 % for Sea-water and 37.4 % for Formation-water.

- The zeta potential measurements support the thesis of Surface charge change as the driving mechanisms of low-salinity effects in limestones. Compared to Formation-water and Sea-water, the Diluted-sea-water and limestone systems exhibited the strongest negative zeta potential. In line with literature and the spontaneous imbibition, centrifuge method and coreflooding results, Surface charge change is hence a reasonable explanation of the wettability alteration in limestones. Moreover, the limestone samples did not show any signs of mineral dissolution.

- The transparent development of a numerical centrifuge and coreflooding model ensures a full understanding of the implemented numerical simulations. The model formulation includes the imposed boundary conditions, fluid properties, hydraulic properties, flow model and material balance coupling. The numerical simulation showed that a simple mathematical formulation is sufficient to history match the experimental data.

- The numerical centrifuge simulation confirmed the analytical centrifuge data analysis. The imbibition capillary pressure curves were characterized by an increasing water-wetting tendency and a simultaneous reduction of the residual oil saturation, as the salinity of the imbibition brines decreased in comparison to Formation-water. In line with the centrifuge results, the numerical coreflooding simulations confirmed a correlation between salinity, wettability and oil recovery. In secondary injection mode, the numerically obtained relative permeability showed the strongest water-wet behavior for Diluted-sea-water, followed by Sea-water and Formation-water.

- In regard to the Middle East reservoirs, the injection of highly diluted Sea-water might be a promising approach to improve oil production. The conducted spontaneous imbibition, centrifuge method and coreflooding tests consistently showed a positive impact on oil recovery. However, since the study was conducted on outcrop samples, it is advisable to validate the obtained results by including reservoir cores.

- The experimental part of the study combined spontaneous imbibition, centrifuge method and USS corefloodings to evaluate the impact of low-salinity effect in limestones samples.

While the centrifuge method shows a clear correlation between the system's salinity and oil recovery, the (centrifuge method) forced imbibition data was furthermore used as the capillary pressure base case in the unsteady coreflooding simulation. However, as described in Chapter 9, the centrifuge method capillary pressure data caused a significant differential pressure overestimation when used in the numerical coreflooding simulations. The discrepancy between the centrifuge method capillary pressure and the numerical capillary pressure might be investigated by the conduction of further experiments such as porous plate method, steady state corefloodings and/or ISSM corefloodings.

Bibliography

[1] Abrams, A., et al.: The influence of fluid viscosity, interfacial tension, and flow velocity on residual oil saturation left by waterflood. Society of Petroleum Engineers Journal **15**(05), 437–447 (1975) 147

[2] Ahmed, T.: Reservoir engineering handbook. Gulf Professional Publishing (2018) 47, 120

[3] Ahr, W.M.: Geology of carbonate reservoirs: the identification, description and characterization of hydrocarbon reservoirs in carbonate rocks. John Wiley & Sons (2011) 4

[4] Al-Attar, H.H., Mahmoud, M.Y., Zekri, A.Y., Almehaideb, R., Ghannam, M.: Low-salinity flooding in a selected carbonate reservoir: experimental approach. Journal of Petroleum Exploration and Production Technology **3**(2), 139–149 (2013) 23, 26, 211

[5] Al Harrasi, A., Al-maamari, R.S., Masalmeh, S.K., et al.: Laboratory investigation of low salinity waterflooding for carbonate reservoirs. In: Abu Dhabi international petroleum conference and exhibition. Society of Petroleum Engineers (2012) 22, 211

[6] Al-Sumaiti, A.M.: Numerical and experimental modeling of double displacement oil recovery in fractured carbonate reservoirs. Ph.D. thesis, Colorado School of Mines (2011) 75

BIBLIOGRAPHY

[7] Allain, O., Dyson, M., Jing, X., Pentland, C., Polikar, M., Suicmez, V.S.: The Imperial College Lectures in Petroleum Engineering: Volume 4: Drilling and Reservoir Appraisal. World Scientific (2018) 78

[8] Alotaibi, M.B., Azmy, R., Nasr-El-Din, H.A., et al.: Wettability challenges in carbonate reservoirs. In: SPE Improved Oil Recovery Symposium. Society of Petroleum Engineers (2010) 28, 52

[9] Alotaibi, M.B., Nasr-El-Din, H.A., Fletcher, J.J., et al.: Electrokinetics of limestone and dolomite rock particles. SPE Reservoir Evaluation & Engineering **14**(05), 594–603 (2011) 26, 27, 29

[10] Anderson, W.G., et al.: Wettability literature survey-part 1: rock/oil/brine interactions and the effects of core handling on wettability. Journal of petroleum technology **38**(10), 1–125 (1986) 27

[11] Anderson, W.G., et al.: Wettability literature survey part 5: the effects of wettability on relative permeability. Journal of Petroleum Technology **39**(11), 1–453 (1987) 53

[12] Asuero, A.G., Sayago, A., Gonzalez, A.: The correlation coefficient: An overview. Critical reviews in analytical chemistry **36**(1), 41–59 (2006) 46

[13] Ayyub, B.M., McCuen, R.H.: Numerical Analysis for Engineers: Methods and Applications. Chapman and Hall/CRC (2015) 113

[14] Aziz, R., Joekar-Niasar, V., Martínez-Ferrer, P.J., Godinez-Brizuela, O.E., Theodoropoulos, C., Mahani, H.: Novel insights into pore-scale dynamics of wettability alteration during low salinity waterflooding. Scientific reports **9**(1), 9257 (2019) 26, 31, 32, 211

BIBLIOGRAPHY

[15] Bahadori, A.: Fluid phase behavior for conventional and unconventional oil and gas reservoirs. Gulf Professional Publishing (2016) 40

[16] Baker, R.O., Yarranton, H.W., Jensen, J.: Practical reservoir engineering and characterization. Gulf Professional Publishing (2015) 50

[17] Basu, S., Ali, M.Y., Farid, A., Berteussen, K.A., Mercado, G.: A microseismic experiment in abu dhabi, united arab emirates: implications for carbonate reservoir monitoring. Arabian Journal of Geosciences **7**(9), 3815–3827 (2014) 14

[18] Bauget, F., Gautier, S., Lenormand, R., Samouillet, A.: Gas-liquid relative permeabilities from one-step and multi-step centrifuge experiments (2012) 126

[19] Bognø, T.: Impacts on oil recovery from capillary pressure and capillary heterogeneities (2008) 75

[20] Bonto, M., Eftekhari, A.A., Nick, H.M.: An overview of the oil-brine interfacial behavior and a new surface complexation model. Scientific reports **9**(1), 6072 (2019) 29

[21] Brady, P.V., Krumhansl, J.L., Mariner, P.E., et al.: Surface complexation modeling for improved oil recovery. In: SPE improved oil recovery symposium. Society of Petroleum Engineers (2012) 29

[22] Chandrasekhar, S., Mohanty, K.K.: Wettability alteration with brine composition in high temperature carbonate reservoirs. In: SPE Annual Technical Conference and Exhibition. Society of Petroleum Engineers (2013) 25, 26, 211

[23] Chang, D., Vinegar, H.J., Morriss, C., Straley, C., et al.: Effective porosity, producible fluid and permeability in carbonates from nmr logging. In: SPWLA 35th Annual Logging Symposium. Society of Petrophysicists and Well-Log Analysts (1994) 49

BIBLIOGRAPHY

[24] Chen, Z.: Reservoir simulation: mathematical techniques in oil recovery, vol. 77. Siam (2007) 117, 136, 138

[25] Chilingar, G.V., Yen, T.: Some notes on wettability and relative permeabilities of carbonate reservoir rocks, ii. Energy Sources **7**(1), 67–75 (1983) 28, 52

[26] Coates, G.R., Galford, J., Mardon, D., Marschall, D., et al.: A new characterization of bulk-volume irreducible using magnetic resonance. The log analyst **39**(01) (1998) 49

[27] Cydarex: Cydar-scal user-manual. user-manual **1**(1) (2018) 121, 125, 126, 142

[28] Dake, L.P.: Fundamentals of reservoir engineering, vol. 8. Elsevier (1983) 18

[29] De Bruin, W.: Simulation of geochemical processes during low salinity water flooding by coupling multiphase buckley-leverett flow to the geochemical package phreeqc (2012) 19, 20

[30] Donaldson, E.C., Ewall, N., Singh, B.: Characteristics of capillary pressure curves. Journal of Petroleum Science and Engineering **6**(3), 249–261 (1991) 78, 81, 86, 121, 142

[31] Donaldson, E.C., Ewall, N., Singh, B.: Characteristics of capillary pressure curves. Journal of Petroleum Science and Engineering **6**(3), 249–261 (1991) 84

[32] Dumux: Dumux handbook 2.12. user-manual **1**(1) (2018) 110, 111, 112, 124

[33] Dumux: Dumux handbook 3.0. user-manual **1**(1) (2019) 115, 116

[34] Feldmann, F., AlSumaiti, A., Masalmeh, S., AlAmeri, W., Oedai, S.: Impact of brine composition and concentration on capillary pressure and residual oil saturation in limestone core samples. In: E3S Web of Conferences, vol. 89, p. 02006. EDP Sciences (2019) 1, 210

BIBLIOGRAPHY

[35] Feldmann, F., Strobel, G., Masalmeh, S., AlSumaiti, A.: An experimental and numerical study of low salinity effects on the oil recovery of carbonate rocks combining spontaneous imbibition, centrifuge method and coreflooding experiments. Journal of Petroleum Science and Engineering p. 107045 (2020) 2, 210

[36] Flemisch, B., Darcis, M., Erbertseder, K., Faigle, B., Lauser, A., Mosthaf, K., Müthing, S., Nuske, P., Tatomir, A., Wolff, M., et al.: Dumux: Dune for multi-{phase, component, scale, physics,...} flow and transport in porous media. Advances in Water Resources **34**(9), 1102–1112 (2011) 110, 111

[37] Forbes, P.: Centrifuge data analysis techniques: An sca survey on the calculation of drainage capillary pressure curves from centrifuge measurements. SCA paper **9714**, 8–10 (1997) 78, 80

[38] Forbes, P.: The h&b boundary condition in centrifuge pc experiments (or why there is not experimental evidence that the pressure field model ever failed). In: International symposium of the society of core analysts, Abu Dhabi, UAE (2000) 77

[39] Forbes, P., et al.: Simple and accurate methods for converting centrifuge data into drainage and imbibition capillary pressure curves. The Log Analyst **35**(04) (1994) 86

[40] Frei, C., Whitney, R., Schiffer, H.W., Rose, K., Rieser, D.A., Al-Qahtani, A., Thomas, P., Turton, H., Densing, M., Panos, E., et al.: World energy scenarios: Composing energy futures to 2050. Tech. rep., Conseil Francais de l'energie (2013) 2, 209

[41] Freire-Gormaly, M., Ellis, J.S., MacLean, H.L., Bazylak, A.: Pore structure characterization of indiana limestone and pink dolomite from pore network reconstructions. Oil & Gas Science and Technology–Revue d'IFP Energies nouvelles **71**(3), 33 (2016) 42, 43

[42] Ghaffari, A., Rahbar-Kelishami, A.: Md simulation and evaluation of the self-diffusion coefficients in aqueous nacl solu-

BIBLIOGRAPHY

tions at different temperatures and concentrations. Journal of Molecular Liquids **187**, 238–245 (2013) 143, 144

[43] Griffiths, D.V., Smith, I.M.: Numerical methods for engineers. Chapman and Hall/CRC (2006) 113

[44] Gupta, R., Maloney, D.R., et al.: Intercept method–a novel technique to correct steady-state relative permeability data for capillary end effects. SPE Reservoir Evaluation & Engineering **19**(02), 316–330 (2016) 95

[45] Gupta, R., Smith, G.G., Hu, L., Willingham, T., Lo Cascio, M., Shyeh, J.J., Harris, C.R., et al.: Enhanced waterflood for carbonate reservoirs-impact of injection water composition. In: SPE Middle East Oil and Gas Show and Conference. Society of Petroleum Engineers (2011) 23, 26, 211

[46] Hadley, G., Handy, L., et al.: A theoretical and experimental study of the steady state capillary end effect. In: Fall Meeting of the Petroleum Branch of AIME. Society of Petroleum Engineers (1956) 95

[47] Hagemann, B.: Numerical and analytical modeling of gas mixing and bio-reactive transport during underground hydrogen storage. Ph.D. thesis, Université de Lorraine (2017) 111, 112, 113

[48] Hassan, T., Wada, Y., et al.: Geology and development of thamama zone 4, zakum field. Journal of Petroleum Technology **33**(07), 1–327 (1981) 3

[49] Hassler, G., Brunner, E., et al.: Measurement of capillary pressures in small core samples. Trans. AIME **160**(1), 114–123 (1945) 76, 78, 79, 80, 118

[50] Heinzl, R.: Concepts for scientific computing. na (2007) 110, 112, 113

BIBLIOGRAPHY

[51] Hiorth, A., Cathles, L., Madland, M.: The impact of pore water chemistry on carbonate surface charge and oil wettability. Transport in porous media **85**(1), 1–21 (2010) 27, 28, 52

[52] Hognesen, E.J., Strand, S., Austad, T., et al.: Waterflooding of preferential oil-wet carbonates: Oil recovery related to reservoir temperature and brine composition. In: SPE Europec/EAGE Annual Conference. Society of Petroleum Engineers (2005) 20

[53] Humphry, K., Suijkerbuijk, B., Van Der Linde, H., Pieterse, S., Masalmeh, S., et al.: Impact of wettability on residual oil saturation and capillary desaturation curves. Petrophysics **55**(04), 313–318 (2014) 77, 96

[54] Jackson, M.D., Al-Mahrouqi, D., Vinogradov, J.: Zeta potential in oil-water-carbonate systems and its impact on oil recovery during controlled salinity water-flooding. Scientific reports **6**, 37,363 (2016) 31

[55] Jadhunandan, P.P.: Effects of brine composition, crude oil, and aging conditions on wettability and oil recovery. Ph.D. thesis, Department of Petroleum Engineering, New Mexico Institute of Mining & Technology (1990) 5

[56] Jerauld, G.R., Webb, K.J., Lin, C.Y., Seccombe, J., et al.: Modeling low-salinity waterflooding. In: SPE Annual Technical Conference and Exhibition. Society of Petroleum Engineers (2006) 20

[57] Jordan, M.M., Sjursaether, K., Collins, I.R., et al.: Scale control within north sea chalk/limestone reservoirs. the challenge of understanding and optimizing chemical placement methods and retention mechanism:-laboratory to field. In: SPE International Symposium and Exhibition on Formation Damage Control. Society of Petroleum Engineers (2004) 17

BIBLIOGRAPHY

[58] Kamath, J., Nakagawa, F., Meyer, R., Kabir, S., Hobbet, R.: Laboratory evaluation of waterflood residual oil saturation in four carbonate cores. Paper SCA **12**, 2001 (2001) 94, 147, 148

[59] Kelleher, H.A., Braun, E.M., Milligan, B.E., Glotzbach, R.C., Haugen, E.: Wettability restoration in cores contaminated by fatty acid emulsifiers. Petrophysics **49**(1), 36 (2008) 43

[60] Lenormand, R., Lorentzen, K., Maas, J.G., Ruth, D., et al.: Comparison of four numerical simulators for scal experiments. Petrophysics **58**(01), 48–56 (2017) 12, 109, 125, 126, 146, 149, 216

[61] Ligthelm, D.J., Gronsveld, J., Hofman, J., Brussee, N., Marcelis, F., van der Linde, H., et al.: Novel waterflooding strategy by manipulation of injection brine composition. In: EUROPEC/EAGE conference and exhibition. Society of Petroleum Engineers (2009) 20, 211

[62] Maas, J., Schulte, A.: Computer simulation of special core analysis (scal) flow experiments shared on the internet. In: SCA-9719 presented at the SCA 1997 conference, Calgary, Canada (1997) 12, 109

[63] Maas, J.G., Flemisch, B., Hebing, A.: Open source simulator dumux available for scal data interpretation. In: SCA, vol. 8, p. 2011 (2011) 12, 117, 149

[64] Mahani, H., Keya, A.L., Berg, S., Bartels, W.B., Nasralla, R., Rossen, W.R.: Insights into the mechanism of wettability alteration by low-salinity flooding (lsf) in carbonates. Energy & Fuels **29**(3), 1352–1367 (2015) 26, 27, 28, 29, 31, 32, 68, 72, 211, 214

[65] Mahani, H., Keya, A.L., Berg, S., Nasralla, R., et al.: The effect of salinity, rock type and ph on the electrokinetics of carbonate-brine interface and surface complexation modeling. In: SPE Reservoir Characterisation and Simulation Conference

BIBLIOGRAPHY

and Exhibition. Society of Petroleum Engineers (2015) 27, 29, 31, 32, 68, 69, 70, 72, 211, 214

[66] Mahani, H., Keya, A.L., Berg, S., Nasralla, R., et al.: Electrokinetics of carbonate/brine interface in low-salinity waterflooding: Effect of brine salinity, composition, rock type, and ph on ζ-potential and a surface-complexation model. Spe Journal **22**(01), 53–68 (2017) 26, 28, 29, 31, 32, 211

[67] Marschall, D., Gardner, J., Mardon, D., Coates, G., et al.: Method for correlating nmr relaxometry and mercury injection data. In: 1995 SCA Conference, paper, 9511 (1995) 50

[68] Masalmeh, S.K.: Impact of capillary forces on residual oil saturation and flooding experiments for mixed to oil-wet carbonate reservoirs. Proceedings of society of core analysts held in Aberdeen, Scotland, UK, SCA2012-11 (2012) 86, 95, 96

[69] Masalmeh, S.K., Sorop, T.G., Suijkerbuijk, B.M., Vermolen, E.C., Douma, S., Van Del Linde, H., Pieterse, S.: Low salinity flooding: Experimental evaluation and numerical interpretation. In: IPTC 2014: International Petroleum Technology Conference (2014) 10, 23, 90

[70] Mascle, M., Youssef, S., Deschamps, H., Vizika, O., et al.: In-situ investigation of aging protocol effect on relative permeability measurements using high-throughput experimentation methods. Petrophysics **60**(04), 514–524 (2019) 54

[71] McGuire, P., Chatham, J., Paskvan, F., Sommer, D., Carini, F., et al.: Low salinity oil recovery: An exciting new eor opportunity for alaska's north slope. In: SPE western regional meeting. Society of Petroleum Engineers (2005) 27

[72] McPhee, C., Reed, J., Zubizarreta, I.: Core analysis: A best practice guide, vol. 64. Elsevier (2015) 9, 11, 41, 43, 45, 49, 50, 75, 77, 78, 80, 93, 96, 128, 147, 160

BIBLIOGRAPHY

[73] Melrose, J., et al.: Interpretation of centrifuge capillary pressure data. The Log Analyst **29**(01) (1988) 77

[74] Meng, Q., Liu, H., Wang, J.: A critical review on fundamental mechanisms of spontaneous imbibition and the impact of boundary condition, fluid viscosity and wettability. Advances in Geo-energy Research **1**(1), 1–17 (2017) 9, 59, 60, 75

[75] Messaâdi, A., Dhouibi, N., Hamda, H., Belgacem, F.B.M., Adbelkader, Y.H., Ouerfelli, N., Hamzaoui, A.H.: A new equation relating the viscosity arrhenius temperature and the activation energy for some newtonian classical solvents. Journal of Chemistry **2015** (2015) 39

[76] Moghaddam, R.N.: A rapid and accurate method for calculation of capillary pressure from centrifuge data. Journal of Petroleum Science and Engineering **135**, 577–582 (2015) 78, 79

[77] Moghaddam, R.N., Jamiolahmady, M.: Steady-state relative permeability measurements of tight and shale rocks considering capillary end effect. Transport in Porous Media **128**(1), 75–96 (2019) 95

[78] Morrow, N.R., Mason, G.: Recovery of oil by spontaneous imbibition. Current Opinion in Colloid & Interface Science **6**(4), 321–337 (2001) 57, 58, 60

[79] Morrow, N.R., et al.: Capillary pressure correlations for uniformly wetted porous media. Journal of Canadian Petroleum Technology **15**(04) (1976) 59, 71

[80] Morrow, N.R., et al.: Wettability and its effect on oil recovery. Journal of petroleum technology **42**(12), 1–476 (1990) 52

[81] Nairn, A., Alsharhan, A.: Sedimentary basins and petroleum geology of the Middle East. Elsevier (1997) 3, 14

BIBLIOGRAPHY

[82] Nasralla, R.A., Mahani, H., van der Linde, H.A., Marcelis, F.H., Masalmeh, S.K., Sergienko, E., Brussee, N.J., Pieterse, S.G., Basu, S.: Low salinity waterflooding for a carbonate reservoir: Experimental evaluation and numerical interpretation. Journal of Petroleum Science and Engineering **164**, 640–654 (2018) 21, 25, 26, 27, 94, 98, 103, 148, 211

[83] Nasralla, R.A., Nasr-El-Din, H.A., et al.: Double-layer expansion: is it a primary mechanism of improved oil recovery by low-salinity waterflooding? SPE Reservoir Evaluation & Engineering **17**(01), 49–59 (2014) 29

[84] OPEC: Annual statistical bulletin 2018 (2018) 2

[85] Park, S.J., Seo, M.K.: Interface science and composites, vol. 18. Academic Press (2011) 30

[86] Pope, G.A., et al.: The application of fractional flow theory to enhanced oil recovery. Society of Petroleum Engineers Journal **20**(03), 191–205 (1980) 19, 20

[87] Qasem, F.H., Nashawi, I.S., Gharbi, R., Mir, M.I.: Recovery performance of partially fractured reservoirs by capillary imbibition. Journal of Petroleum Science and Engineering **60**(1), 39–50 (2008) 60

[88] Rapp, B.E.: Microfluidics: Modeling, Mechanics and Mathematics. William Andrew (2016) 111, 112

[89] Romanuka, J., Hofman, J., Ligthelm, D.J., Suijkerbuijk, B., Marcelis, F., Oedai, S., Brussee, N., van der Linde, H., Aksulu, H., Austad, T., et al.: Low salinity eor in carbonates. In: SPE Improved Oil Recovery Symposium. Society of Petroleum Engineers (2012) 20, 21, 22, 26, 27, 91, 211

[90] Selvamani, V.: Stability studies on nanomaterials used in drugs. In: Characterization and Biology of Nanomaterials for Drug Delivery, pp. 425–444. Elsevier (2019) 29

BIBLIOGRAPHY

[91] Shehata, A.M., Alotaibi, M.B., Nasr-El-Din, H.A., et al.: Waterflooding in carbonate reservoirs: does the salinity matter? SPE Reservoir Evaluation & Engineering **17**(03), 304–313 (2014) 23, 26, 211

[92] Sheng, J.: Modern chemical enhanced oil recovery: theory and practice. Gulf Professional Publishing (2010) 3

[93] Sheng, J.: Enhanced oil recovery field case studies. Gulf Professional Publishing (2013) 4

[94] Sheng, J.: Critical review of low-salinity waterflooding. Journal of Petroleum Science and Engineering **120**, 216–224 (2014) 5, 17, 27, 57

[95] Sheng, J.J.: Review of surfactant enhanced oil recovery in carbonate reservoirs. Advances in Petroleum Exploration and Development **6**(1), 1–10 (2013) 2, 4, 14

[96] Song, J., Zeng, Y., Wang, L., Duan, X., Puerto, M., Chapman, W.G., Biswal, S.L., Hirasaki, G.J.: Surface complexation modeling of calcite zeta potential measurements in brines with mixed potential determining ions ($ca2+$, $co32-$, $mg2+$, $so42-$) for characterizing carbonate wettability. Journal of colloid and interface science **506**, 169–179 (2017) 26, 29, 31

[97] Sorop, T.G., Suijkerbuijk, B.M., Masalmeh, S.K., Looijer, M.T., Parker, A.R., Dindoruk, D.M., Goodyear, S.G., Al-Qarshubi, I.S., et al.: Integrated approach in deploying low salinity waterflooding. In: SPE Enhanced Oil Recovery Conference. Society of Petroleum Engineers (2013) 95

[98] Strand, S., Austad, T., Puntervold, T., Høgnesen, E.J., Olsen, M., Barstad, S.M.F.: "smart water" for oil recovery from fractured limestone: a preliminary study. Energy & fuels **22**(5), 3126–3133 (2008) 17, 18, 20

[99] Tetteh, J., Janjang, N.M., Barati, R., et al.: Wettability alteration and enhanced oil recovery using low salinity waterflooding in limestone rocks: A mechanistic study. In: SPE Kingdom

of Saudi Arabia Annual Technical Symposium and Exhibition. Society of Petroleum Engineers (2018) 24, 26, 27, 211

[100] TheGulfIntelligence.com: A gulf intelligence industry survey report (2015) 2

[101] Tiab, D., Donaldson, E.C.: Petrophysics: theory and practice of measuring reservoir rock and fluid transport properties. Gulf professional publishing (2015) 58, 79, 80

[102] United Nations, D.o.E., Affairs, S.: World population prospects 2019: Highlights (2019) 2

[103] Van Cappellen, P., Charlet, L., Stumm, W., Wersin, P.: A surface complexation model of the carbonate mineral-aqueous solution interface. Geochimica et Cosmochimica Acta **57**(15), 3505–3518 (1993) 29

[104] Vega, B., Kovscek, A.: Carbon dioxide (co2) sequestration in oil and gas reservoirs and use for enhanced oil recovery (eor). In: Developments and Innovation in Carbon Dioxide (CO2) Capture and Storage Technology, pp. 104–126. Elsevier (2010) 41

[105] Warren, J., Root, P.J., et al.: The behavior of naturally fractured reservoirs. Society of Petroleum Engineers Journal **3**(03), 245–255 (1963) 59

[106] Winoto, W., Loahardjo, N., Xie, S.X., Yin, P., Morrow, N.R., et al.: Secondary and tertiary recovery of crude oil from outcrop and reservoir rocks by low salinity waterflooding. In: SPE Improved Oil Recovery Symposium. Society of Petroleum Engineers (2012) 24, 26, 211

[107] Yi, Z., Sarma, H.K., et al.: Improving waterflood recovery efficiency in carbonate reservoirs through salinity variations and ionic exchanges: A promising low-costßmartwaterfloodäpproach. In: Abu Dhabi International Petroleum

BIBLIOGRAPHY

Conference and Exhibition. Society of Petroleum Engineers (2012) 22, 211

[108] Yousef, A.A., Al-Saleh, S., Al-Kaabi, A.U., Al-Jawfi, M.S., et al.: Laboratory investigation of novel oil recovery method for carbonate reservoirs. In: Canadian Unconventional Resources and International Petroleum Conference. Society of Petroleum Engineers (2010) 24, 26, 211

[109] Yousef, A.A., Al-Saleh, S.H., Al-Kaabi, A., Al-Jawfi, M.S., et al.: Laboratory investigation of the impact of injection-water salinity and ionic content on oil recovery from carbonate reservoirs. SPE Reservoir Evaluation & Engineering **14**(05), 578–593 (2011) 24, 26, 27, 211

[110] Zaretskiy, Y., et al.: Towards modelling physical and chemical effects during wettability alteration in carbonates at pore and continuum scales. Ph.D. thesis, Heriot-Watt University (2012) 27

[111] Zhang, P., Tweheyo, M.T., Austad, T.: Wettability alteration and improved oil recovery by spontaneous imbibition of seawater into chalk: Impact of the potential determining ions Ca^{2+}, Mg^{2+}, and SO_4^{2-}. Colloids and Surfaces A: Physicochemical and Engineering Aspects **301**(1-3), 199–208 (2007) 17, 27, 29, 30, 70, 72, 214

Appendix A

Experimental study appendix

A.1 Spontaneous and forced imbibition results

The results of the spontaneous and centrifuge method experiments are listed in Appendix A.1, Table A.1 to Table A.14. Furthermore, sample properties such as core length, diameter and pore volume are summarized. In addition, the section enumerates the results of the absolute brine permeability measurements and effective oil permeability measurements before and after aging.

A.2 Unsteady state coreflooding results

The results of the three unsteady coreflooding are summarized in Appendix A.2, Table A.15 to Table A.17. Moreover, the fundamental core properties and the absolute/effective permeability measurements listed.

A.2 UNSTEADY STATE COREFLOODING RESULTS

Sample	In1	Length [mm]	48.59	Sw connate	0.302
CW	FW	Diameter [mm]	37.93	Sw spon. imb.	0.554
IW	DSW	P_V [cm^3]	8.56	Sw forced imb.	0.919

	Flowrate number	Flowrate [cm^3/min]	ΔP [barg]	Corr. ΔP [barg]	Perm. [md]
Absolute brine perm.	1	0.2	0.39	0.24	8.83
	2	0.5	0.80	0.64	8.21
	3	1	1.40	1.24	8.49
	linear coef.	0.9991	offset	0.15	$\bar{k} = \mathbf{8.51}$
Effective oil perm. before aging	1	0.20	0.48	0.42	9.15
	2	0.50	1.07	1.01	9.56
	3	1.00	2.12	2.05	9.37
	linear coef.	0.9997	offset	0.06	$\bar{k} = \mathbf{9.36}$
Effective oil perm. after aging	1	0.5	1.42	1.37	7.00
	2	1	2.77	2.72	7.07
	3	1.5	4.16	4.11	7.02
	linear coef.	0.9999	offset	0.05	$\bar{k} = \mathbf{7.03}$

Spontaneous imbibition		Centrifuge method			
Time [h]	Sw	Time [h]	RPM	Inlet P [barg]	Sw
1	0.302	0	0	0.00	0.554
24	0.332	30	1000	-0.14	0.634
48	0.367	60	1280	-0.22	0.656
120	0.426	90	1650	-0.37	0.685
168	0.461	120	2110	-0.60	0.725
216	0.490	155	2700	-0.99	0.772
312	0.507	185	3470	-1.63	0.822
360	0.536	215	4450	-2.68	0.863
432	0.554	245	5700	-4.40	0.895
480	0.554	275	7500	-7.61	0.919

Table A.1: Spontaneous and forced imbibition results In1.

A.2 UNSTEADY STATE COREFLOODING RESULTS

Sample	In2	Length [mm]	48.68	Sw connate	0.264
CW	FW	Diameter [mm]	37.95	Sw spon. imb.	0.451
IW	SW	P_V [cm^3]	8.60	Sw forced imb.	0.883

		Flowrate number	Flowrate [cm^3/min]	ΔP [$barg$]	Corr. ΔP [$barg$]	Perm. [md]
Absolute brine perm.		1	0.2	0.13	0.14	14.99
		2	0.5	0.34	0.35	15.05
		3	1.0	0.69	0.70	15.02
		linear coef.	0.9999	offset	-0.01	$\bar{k} = 15.01$
Effective oil perm. before aging		1	0.5	0.88	0.72	12.68
		2	1.0	1.63	1.47	12.75
		3	1.5	2.33	2.17	13.05
		linear coef.	0.9996	offset	0.16	$\bar{k} = 12.82$
Effective oil perm. after aging		1	0.5	0.90	0.89	10.98
		2	1	1.77	1.76	11.00
		3	1.5	2.62	2.61	11.12
		linear coef.	0.9999	offset	0.01	$\bar{k} = 11.13$

Spontaneous imbibition		Centrifuge method			
Time [h]	Sw	Time [h]	RPM	Inlet P [$barg$]	Sw
1	0.264	0	0	0.00	0.451
24	0.277	30	1000	-0.16	0.583
48	0.294	60	1280	-0.27	0.614
120	0.341	90	1650	-0.45	0.648
168	0.376	120	2110	-0.73	0.702
216	0.411	155	2700	-1.19	0.748
312	0.428	185	3470	-1.97	0.792
360	0.434	215	4450	-3.24	0.830
432	0.451	245	5700	-5.31	0.859
480	0.451	275	7500	-9.20	0.883

Table A.2: Spontaneous and forced imbibition results In2.

A.2 UNSTEADY STATE COREFLOODING RESULTS

Sample	In3	Length [mm]	48.70	Sw connate	0.270
CW	FW	Diameter [mm]	37.96	Sw spon. imb.	0.284
IW	FW	P_V [cm^3]	8.28	Sw forced imb.	0.805

	Flowrate number	Flowrate [cm^3/min]	ΔP [barg]	Corr. ΔP [barg]	Perm. [md]
Absolute brine perm.	1	0.2	0.27	0.27	8.75
	2	0.5	0.66	0.66	8.40
	3	1.0	1.26	1.26	8.56
	linear coef.	0.9996	offset	0.03	$\bar{k} = 8.57$
Effective oil perm. before aging	1	0.5	1.00	1.04	9.29
	2	1	2.07	2.10	9.15
	3	1.5	3.12	3.16	9.14
	linear coef.	0.9999	offset	-0.04	$\bar{k} = 9.12$
Effective oil perm. after aging	1	0.5	1.27	1.31	7.29
	2	1.0	2.59	2.64	7.29
	3	1.5	3.92	3.97	7.27
	linear coef.	1.0000	offset	-0.05	$\bar{k} = 7.27$

Spontaneous imbibition		Centrifuge method			
Time [h]	Sw	Time [h]	RPM	Inlet P [barg]	Sw
1	0.270	0	0	0.00	0.284
24	0.270	30	1000	-0.23	0.445
48	0.270	60	1280	-0.38	0.484
120	0.278	90	1650	-0.63	0.533
168	0.284	120	2110	-1.04	0.597
216	0.284	150	2700	-1.70	0.659
312	0.284	180	3470	-2.80	0.712
360	0.284	204	4450	-4.61	0.748
432	0.284	228	5700	-7.56	0.778
480	0.284	252	7500	-13.09	0.805

Table A.3: Spontaneous and forced imbibition results In3.

A.2 UNSTEADY STATE COREFLOODING RESULTS

Sample	In4	Length [mm]	48.22	Sw connate	0.312
CW	DSW	Diameter [mm]	37.93	Sw spon. imb.	0.333
IW	DSW	P_V [cm^3]	8.11	Sw forced imb.	0.925

	Flowrate number	Flowrate [cm^3/min]	ΔP [$barg$]	Corr. ΔP [$barg$]	Perm. [md]
Absolute brine perm.	1	0.2	0.23	0.21	6.73
	2	0.5	0.56	0.54	6.58
	3	1.0	1.09	1.07	6.65
	linear coef.	0.9999	offset	0.02	$\bar{k} = 6.65$
Effective oil perm. before aging	1	0.2	0.69	0.72	5.32
	2	0.5	1.71	1.73	5.52
	3	1.0	3.48	3.51	5.44
	linear coef.	0.9999	offset	-0.02	$\bar{k} = 5.45$
Effective oil perm. after aging	1	0.2	1.09	1.06	3.89
	2	0.5	2.66	2.63	3.75
	3	1	5.30	5.27	3.69
	linear coef.	0.9995	offset	0.03	$\bar{k} = 3.78$

Spontaneous imbibition		Centrifuge method			
Time [h]	Sw	Time [h]	RPM	Inlet P [$barg$]	Sw
1	0.312	0	0	0.00	0.333
24	0.312	30	1000	-0.13	0.461
48	0.320	60	1280	-0.22	0.516
120	0.320	90	1650	-0.37	0.579
168	0.320	120	2110	-0.60	0.636
216	0.320	150	2700	-0.98	0.709
312	0.333	180	3470	-1.62	0.782
360	0.333	210	4450	-2.66	0.846
432	0.333	240	5700	-4.37	0.894
480	0.333	265	7500	-7.56	0.925

Table A.4: Spontaneous and forced imbibition results In4.

A.2 UNSTEADY STATE COREFLOODING RESULTS

Sample	In5	Length [mm]	48.31	Sw connate	0.245
CW	FW	Diameter [mm]	37.91	Sw spon. imb.	0.252
IW	FW	P_V [cm^3]	8.32	Sw forced imb.	0.825

		Flowrate number	Flowrate [cm^3/min]	ΔP [barg]	Corr. ΔP [barg]	Perm. [md]
Absolute brine perm.		1	0.2	0.24	0.19	10.88
		2	0.5	0.54	0.49	10.60
		3	1.0	1.02	0.98	10.73
		linear coef.	0.9999	offset	0.04	$\overline{k} = 10.74$
Effective oil perm. before aging		1	0.5	0.79	0.79	12.56
		2	1	1.54	1.54	12.65
		3	1.5	2.38	2.37	12.23
		linear coef.	0.9994	offset	0.01	$\overline{k} = 12.55$
Effective oil perm. after aging		1	0.5	1.23	1.22	7.91
		2	1	2.45	2.44	7.89
		3	1.5	3.64	3.63	7.94
		linear coef.	1.0000	offset	0.01	$\overline{k} = 7.91$

Spontaneous imbibition		Centrifuge method			
Time [h]	Sw	Time [h]	RPM	Inlet P [barg]	Sw
1	0.245	0	0	0.00	0.252
24	0.245	30	1000	-0.23	0.412
48	0.245	60	1280	-0.38	0.463
120	0.245	90	1650	-0.63	0.517
168	0.252	120	2110	-1.03	0.585
216	0.252	150	2700	-1.68	0.653
312	0.252	180	3470	-2.78	0.714
360	0.252	204	4450	-4.58	0.759
432	0.252	228	5700	-7.51	0.796
480	0.252	252	7500	-13.00	0.825

Table A.5: Spontaneous and forced imbibition results In5.

A.2 UNSTEADY STATE COREFLOODING RESULTS

Sample	In7	Length [mm]	48.42	Sw connate	0.324
CW	SW	Diameter [mm]	37.84	Sw spon. imb.	0.340
IW	SW	P_V [cm^3]	7.77	Sw forced imb.	0.891

	Flowrate number	Flowrate [cm^3/min]	ΔP [$barg$]	Corr. ΔP [$barg$]	Perm. [md]
Absolute brine perm.	1	0.2	0.51	0.40	3.98
	2	0.5	1.08	1.09	4.07
	3	1	2.07	2.08	4.03
	linear coef.	0.9999	offset	0.11	$\bar{k} = 4.03$
Effective oil perm. before aging	1	0.5	2.32	2.77	3.75
	2	1	4.99	5.44	3.68
	3	1.5	7.81	8.26	3.59
	linear coef.	0.9988	offset	-0.45	$\bar{k} = 3.38$
Effective oil perm. after aging	1	0.5	4.25	3.77	2.55
	2	1.0	7.99	7.51	2.57
	3	1.5	12.37	11.89	2.43
	linear coef.	1.0000	offset	0.48	$\bar{k} = 2.57$

Spontaneous imbibition		Centrifuge method			
Time [h]	Sw	Time [h]	RPM	Inlet P [$barg$]	Sw
1	0.324	0	0	0.00	0.340
24	0.324	30	1000	-0.16	0.526
48	0.334	60	1280	-0.27	0.570
120	0.334	90	1650	-0.44	0.626
168	0.334	120	2110	-0.72	0.685
216	0.334	150	2700	-1.19	0.744
312	0.334	180	3470	-1.96	0.794
360	0.334	210	4450	-3.22	0.831
432	0.340	240	5700	-5.29	0.862
480	0.340	270	7500	-9.15	0.891

Table A.6: Spontaneous and forced imbibition results In7.

A.2 UNSTEADY STATE COREFLOODING RESULTS

Sample	In9	Length [mm]	48.43	Sw connate	0.257
CW	FW	Diameter [mm]	37.93	Sw spon. imb.	0.515
IW	DSW	P_V [cm^3]	8.39	Sw forced imb.	0.902

	Flowrate number	Flowrate [cm^3/min]	ΔP [$barg$]	Corr. ΔP [$barg$]	Perm. [md]
Absolute brine perm.	1	0.2	0.17	0.18	11.87
	2	0.5	0.45	0.46	11.47
	1	0.89	0.90	11.65	
	linear coef.	0.9998	offset	-0.01	$\bar{k} = 11.66$
Effective oil perm. before aging	1	0.5	0.76	0.84	11.46
	2	1	1.59	1.66	11.52
	3	1.5	2.43	2.51	11.48
	linear coef.	1.0000	offset	-0.08	$\bar{k} = 11.49$
Effective oil perm. after aging	1	0.5	1.19	1.19	8.08
	2	1	2.36	2.36	8.12
	3	1.5	3.54	3.54	8.12
	linear coef.	1.0000	offset	0.00	$\bar{k} = 8.13$

Spontaneous imbibition		Centrifuge method			
Time [h]	Sw	Time [h]	RPM	Inlet P [$barg$]	Sw
1	0.257	0	0	0.00	0.515
24	0.277	30	1000	-0.13	0.603
48	0.301	60	1280	-0.22	0.626
120	0.366	90	1650	-0.37	0.656
168	0.408	120	2110	-0.60	0.698
216	0.438	155	2700	-0.98	0.747
312	0.461	185	3470	-1.62	0.800
360	0.491	215	4450	-2.67	0.843
432	0.515	245	5700	-4.38	0.876
480	0.515	275	7500	-7.59	0.902

Table A.7: Spontaneous and forced imbibition results In9.

A.2 UNSTEADY STATE COREFLOODING RESULTS

Sample	In10	Length [mm]	48.69	Sw connate	0.304
CW	FW	Diameter [mm]	37.94	Sw spon. imb.	0.434
IW	SW	P_V [cm^3]	8.21	Sw forced imb.	0.850

	Flowrate number	Flowrate [cm^3/min]	ΔP [$barg$]	Corr. ΔP [$barg$]	Perm. [md]
Absolute brine perm.	1	0.2	0.35	0.34	6.32
	2	0.5	0.88	0.87	6.10
	3	1.0	1.73	1.72	6.15
	linear coef.	0.9999	offset	0.01	$\overline{k}=\mathbf{6.19}$
Effective oil perm. before aging	1	0.5	1.75	1.73	5.57
	2	1	3.40	3.38	5.69
	3	1.5	5.18	5.16	5.60
	linear coef.	0.9998	offset	0.02	$\overline{k}=\mathbf{5.61}$
Effective oil perm. after aging	1	0.2	1.94	1.65	2.70
	2	0.5	4.39	4.11	2.48
	3	1	8.52	8.24	2.41
	linear coef.	1.0000	offset	0.28	$\overline{k}=\mathbf{2.53}$

Spontaneous imbibition		Centrifuge method			
Time [h]	Sw	Time [h]	RPM	Inlet P [$barg$]	Sw
1	0.304	0	0	0.00	0.434
24	0.304	30	1000	-0.16	0.595
48	0.349	60	1280	-0.27	0.630
120	0.349	90	1650	-0.45	0.659
168	0.367	120	2110	-0.73	0.701
216	0.379	150	2700	-1.19	0.741
312	0.404	180	3470	-1.97	0.778
360	0.422	210	4450	-3.24	0.803
432	0.434	240	5700	-5.31	0.828
480	0.434	270	7500	-9.20	0.850

Table A.8: Spontaneous and forced imbibition results In10.

A.2 UNSTEADY STATE COREFLOODING RESULTS

Sample	In12	Length [mm]	48.61	Sw connate	0.329
CW	SW	Diameter [mm]	37.84	Sw spon. imb.	0.343
IW	SW	P_V [cm^3]	8.30	Sw forced imb.	0.906

		Flowrate number	Flowrate [cm^3/min]	ΔP [$barg$]	Corr. ΔP [$barg$]	Perm. [md]
Absolute brine perm.		1	0.2	0.23	0.19	8.23
		2	0.5	0.56	0.52	7.54
		3	1	1.05	1.01	7.84
		linear coef.	0.9987	offset	0.04	$\bar{k} = 7.87$
Effective oil perm. before aging		1	0.2	0.51	0.55	7.01
		2	0.5	1.30	1.34	7.22
		3	1	2.68	2.72	7.12
		linear coef.	0.9998	offset	-0.04	$\bar{k} = 7.12$
Effective oil perm. after aging		1	0.2	0.94	0.72	5.46
		2	1.0	4.21	3.99	4.86
		3	1.5	6.13	5.91	4.92
		linear coef.	0.9994	offset	0.22	$\bar{k} = 4.87$

Spontaneous imbibition		Centrifuge method			
Time [h]	Sw	Time [h]	RPM	Inlet P [$barg$]	Sw
1	0.329	0	0	0.00	0.343
24	0.329	30	1000	-0.16	0.584
84	0.331	60	1280	-0.27	0.638
108	0.331	90	1650	-0.44	0.684
156	0.331	120	2110	-0.73	0.733
180	0.331	150	2700	-1.19	0.784
276	0.331	180	3470	-1.97	0.825
324	0.331	210	4450	-3.23	0.856
624	0.343	240	5700	-5.30	0.882
-	-	270	7500	-9.18	0.906

Table A.9: Spontaneous and forced imbibition results In12.

A.2 UNSTEADY STATE COREFLOODING RESULTS

Sample	In13	Length [mm]	49.08	Sw connate	0.298
CW	DSW	Diameter [mm]	37.84	Sw spon. imb.	0.318
IW	DSW	P_V [cm^3]	8.51	Sw forced imb.	0.923

		Flowrate number	Flowrate [cm^3/min]	ΔP [barg]	Corr. ΔP [barg]	Perm. [md]
Absolute brine perm.		1	0.50	0.30	0.26	13.78
		2	1.00	0.56	0.53	14.88
		3	1.50	0.83	0.79	14.93
		linear coef.	1.0000	offset	0.03	$\bar{k} = 14.53$
Effective oil perm. before aging		1	0.5	0.94	0.98	9.96
		2	1	1.95	1.99	9.82
		3	1.5	2.92	2.95	9.92
		linear coef.	0.9998	offset	-0.04	$\bar{k} = 9.90$
Effective oil perm. after aging		1	0.5	1.35	1.41	6.91
		2	1.0	2.71	2.77	7.05
		3	1.5	4.15	4.21	6.95
		linear coef.	0.9997	offset	-0.06	$\bar{k} = 6.97$

Spontaneous imbibition		Centrifuge method			
Time [h]	Sw	Time [h]	RPM	Inlet P [barg]	Sw
1	0.298	0	0	0.00	0.318
24	0.300	30	1000	-0.14	0.486
84	0.306	60	1280	-0.22	0.551
108	0.312	90	1650	-0.37	0.608
156	0.312	120	2110	-0.60	0.661
180	0.318	150	2700	-0.99	0.728
276	0.318	180	3470	-1.63	0.797
324	0.318	210	4450	-2.69	0.847
624	0.318	240	5700	-4.41	0.887
-	-	270	7500	-7.63	0.923

Table A.10: Spontaneous and forced imbibition results In13.

A.2 UNSTEADY STATE COREFLOODING RESULTS

Sample	In14	Length [mm]	45.16	Sw connate	0.240
CW	SW	Diameter [mm]	37.84	Sw spon. imb.	0.240
IW	FW	P_V [cm^3]	7.79	Sw forced imb.	0.802

	Flowrate number	Flowrate [cm^3/min]	ΔP [$barg$]	Corr. ΔP [$barg$]	Perm. [md]
Absolute brine perm.	1	0.5	0.31	0.28	11.96
	2	1	0.61	0.58	11.56
	3	1.5	0.88	0.85	11.82
	linear coef.	0.9991	offset	0.03	$\bar{k} = \mathbf{11.78}$
Effective oil perm. before aging	1	0.5	0.58	0.77	11.24
	2	1	1.40	1.59	11.09
	3	1.5	2.14	2.33	11.41
	linear coef.	0.9993	offset	-0.19	$\bar{k} = \mathbf{11.22}$
Effective oil perm. after aging	1	0.5	1.07	1.01	8.86
	2	1	2.03	1.98	9.06
	3	1.5	3.07	3.02	8.93
	linear coef.	0.9996	offset	0.05	$\bar{k} = \mathbf{8.95}$

Spontaneous imbibition		Centrifuge method			
Time [h]	Sw	Time [h]	RPM	Inlet P [$barg$]	Sw
1	0.240	0	0	0.00	0.240
24	0.240	30	1000	-0.22	0.403
84	0.240	60	1280	-0.36	0.458
108	0.240	90	1650	-0.59	0.502
156	0.240	120	2110	-0.97	0.569
180	0.240	150	2700	-1.59	0.645
276	0.240	180	3470	-2.62	0.702
324	0.240	210	4450	-4.31	0.741
624	0.240	240	5700	-7.08	0.778
-	-	270	7500	-12.25	0.802

Table A.11: Spontaneous and forced imbibition results In14.

A.2 UNSTEADY STATE COREFLOODING RESULTS

Sample	In15	Length [mm]	44.86	Sw connate	0.263
CW	DSW	Diameter [mm]	37.77	Sw spon. imb.	0.263
IW	SW	P_V [cm^3]	7.47	Sw forced imb.	0.857

		Flowrate number	Flowrate [cm^3/min]	ΔP [$barg$]	Corr. ΔP [$barg$]	Perm. [md]
Absolute brine perm.		1	0.5	0.77	0.65	5.12
		2	1	1.45	1.33	5.01
		3	1.5	2.09	1.97	5.08
		linear coef.	0.9996	offset	0.12	$\bar{k} = 5.07$
Effective oil perm. before aging		1	0.5	1.76	1.65	5.44
		2	1	3.34	3.23	5.55
		3	1.5	5.02	4.91	5.47
		linear coef.	0.9997	offset	0.11	$\bar{k} = 5.49$
Effective oil perm. after aging		1	0.5	1.83	1.84	4.87
		2	1	3.68	3.69	4.86
		3	1.5	5.51	5.52	4.87
		linear coef.	1.0000	offset	-0.01	$\bar{k} = 4.87$

Spontaneous imbibition		Centrifuge method			
Time [h]	Sw	Time [h]	RPM	Inlet P [$barg$]	Sw
1	0.263	0	0	0.00	0.263
24	0.263	30	1000	-0.15	0.392
84	0.263	60	1280	-0.25	0.455
108	0.263	90	1650	-0.41	0.511
156	0.263	120	2110	-0.68	0.590
180	0.263	150	2700	-1.11	0.674
276	0.263	180	3470	-1.83	0.741
324	0.263	210	4450	-3.01	0.789
624	0.263	240	5700	-4.93	0.830
-	-	270	7500	-8.54	0.857

Table A.12: Spontaneous and forced imbibition results In15.

A.2 UNSTEADY STATE COREFLOODING RESULTS

Sample	In16	Length [mm]	45.96	Sw connate	0.286
CW	SW	Diameter [mm]	37.85	Sw spon. imb.	0.446
IW	DSW	P_V [cm^3]	7.42	Sw forced imb.	0.852

	Flowrate number	Flowrate [cm^3/min]	ΔP [barg]	Corr. ΔP [barg]	Perm. [md]
Absolute brine perm.	1	1	1.68	1.57	4.76
	2	1.5	2.45	2.33	4.81
	3	2	3.25	3.13	4.77
	linear coef.	0.9998	offset	0.12	$\bar{k}=\mathbf{4.78}$
Effective oil perm. before aging	1	0.5	1.51	1.42	6.42
	2	1	2.95	2.86	6.39
	3	1.5	4.36	4.28	6.41
	linear coef.	1.0000	offset	0.09	$\bar{k}=\mathbf{6.41}$
Effective oil perm. after aging	1	0.5	1.59	1.65	5.55
	2	1	3.14	3.20	5.72
	3	1.5	4.83	4.89	5.60
	linear coef.	0.9993	offset	-0.06	$\bar{k}=\mathbf{5.62}$

Spontaneous imbibition		Centrifuge method			
Time [h]	Sw	Time [h]	RPM	Inlet P [barg]	Sw
1	0.286	0	0	0.00	0.446
24	0.298	30	1000	-0.13	0.543
84	0.318	60	1280	-0.21	0.586
108	0.325	90	1650	-0.35	0.624
156	0.352	120	2110	-0.57	0.672
180	0.366	150	2700	-0.94	0.714
276	0.399	180	3470	-1.55	0.759
324	0.419	210	4450	-2.54	0.793
624	0.446	240	5700	-4.17	0.824
-	-	270	7500	-7.22	0.852

Table A.13: Spontaneous and forced imbibition results In16.

A.2 UNSTEADY STATE COREFLOODING RESULTS

Sample	In17	Length [mm]	46.11	Sw connate	0.258
CW	SW	Diameter [mm]	37.86	Sw spon. imb.	0.424
IW	DSW	P_V [cm^3]	7.75	Sw forced imb.	0.874

	Flowrate number	Flowrate [cm^3/min]	ΔP [$barg$]	Corr. ΔP [$barg$]	Perm. [md]
Absolute brine perm.	1	1	0.62	0.52	14.50
	2	1.5	0.88	0.78	14.48
	3	2	1.14	1.03	14.50
	linear coef.	1.0000	offset	0.11	$\overline{k} = 14.49$
Effective oil perm. before aging	1	0.5	0.95	0.81	11.33
	2	1	1.81	1.66	11.02
	3	1.5	2.59	2.45	11.22
	linear coef.	0.9994	offset	0.14	$\overline{k} = 11.19$
Effective oil perm. after aging	1	0.5	0.94	0.81	9.78
	2	1	1.79	1.66	10.24
	3	1.5	2.57	2.44	10.68
	linear coef.	0.9994	offset	0.13	$\overline{k} = 10.23$

Spontaneous imbibition		Centrifuge method			
Time [h]	Sw	Time [h]	RPM	Inlet P [$barg$]	Sw
1	0.258	0	0	0.00	0.424
24	0.269	30	1000	-0.13	0.570
84	0.282	60	1280	-0.21	0.606
108	0.295	90	1650	-0.35	0.642
156	0.321	120	2110	-0.57	0.684
180	0.333	150	2700	-0.94	0.734
276	0.366	180	3470	-1.55	0.777
324	0.385	210	4450	-2.55	0.814
624	0.424	240	5700	-4.18	0.846
-	-	270	7500	-7.24	0.874

Table A.14: Spontaneous and forced imbibition results In17.

A.2 UNSTEADY STATE COREFLOODING RESULTS

Sample	In2b	Length [mm]	48.68	Sw connate	0.218
P_V [cm^3]	8.45	Diameter [mm]	37.95	Sw USS	0.626

	Flowrate number	Flowrate [cm^3/min]	ΔP [$barg$]	Corr. ΔP [$barg$]	Perm. [md]
Absolute brine perm.	1	1	0.68	0.70	14.98
	2	1.5	1.03	1.05	15.01
	3	2	1.39	1.42	14.89
	linear coef.	0.9999	offset	-0.04	$\bar{k} = 14.97$
Effective oil perm. before aging	1	1	1.86	1.71	11.90
	2	1.5	2.69	2.54	11.80
	3	2	3.48	3.34	11.88
	linear coef.	0.9998	offset	0.24	$\bar{k} = 11.86$
Effective oil perm. after aging	1	0.5	1.07	0.94	9.94
	2	1	2.05	1.92	9.91
	3	1.5	3.01	2.88	9.93
	linear coef.	1.0000	offset	0.10	$\bar{k} = 9.93$

USS coreflooding experiments					
FW injection		SW injection		DSW injection	
Inj. rate [cm^3/min]	Sw	Inj. rate [cm^3/min]	Sw	Inj. rate [cm^3/min]	Sw
0.05	0.390	0.05	0.626	0.05	0.626
0.2	0.496	0.2	0.626	0.2	0.626
1	0.591	1	0.626	1	0.626
3	0.609	3	0.626	3	0.626
5	0.626	5	0.626	5	0.626

Table A.15: USS coreflooding results In2b.

A.2 UNSTEADY STATE COREFLOODING RESULTS

Sample	In4b	Length [mm]	48.68	Sw connate	0.265
P_V [cm^3]	7.90	Diameter [mm]	37.95	Sw USS	0.683

	Flowrate number	Flowrate [cm^3/min]	ΔP [$barg$]	Corr. ΔP [$barg$]	Perm. [md]
Absolute brine perm.	1	1	1.40	1.41	7.42
	2	1.5	2.10	2.11	7.42
	3	2	2.81	2.82	7.42
	linear coef.	1.0000	offset	-0.01	$\overline{k} = \mathbf{7.41}$
Effective oil perm. before aging	1	1	2.53	2.38	7.24
	2	1.5	3.87	3.73	7.20
	3	2	5.17	5.03	7.23
	linear coef.	0.9999	offset	-0.11	$\overline{k} = \mathbf{7.22}$
Effective oil perm. after aging	1	0.5	2.14	2.01	4.54
	2	1	4.06	3.93	4.75
	3	1.5	6.25	6.12	4.61
	linear coef.	0.9985	offset	0.04	$\overline{k} = \mathbf{4.63}$

USS coreflooding experiments			
Sw injection		DSW injection	
Inj. rate [cm^3/min]	Sw	Inj. rate [cm^3/min]	Sw
0.05	0.487	0.05	0.645
0.2	0.575	0.2	0.645
1	0.619	1	0.670
3	0.645	3	0.676
5	0.645	5	0.683

Table A.16: USS coreflooding results In4b.

A.2 UNSTEADY STATE COREFLOODING RESULTS

Sample	In9b	Length [mm]	48.68	Sw connate	0.249
P_V [cm^3]	8.31	Diameter [mm]	37.95	Sw USS	0.698

		Flowrate number	Flowrate [cm^3/min]	ΔP [$barg$]	Corr. ΔP [$barg$]	Perm. [md]
Absolute brine perm.		1	1	0.82	0.85	12.33
		2	1.5	1.24	1.28	12.30
		3	2	1.67	1.70	12.32
		linear coef.	1.0000	offset	-0.04	$\bar{k} = \mathbf{12.32}$
Effective oil perm. before aging		1	1	2.15	2.00	9.75
		2	1.5	3.13	2.98	9.75
		3	2	4.11	3.97	9.75
		linear coef.	1.0000	offset	0.18	$\bar{k} = \mathbf{9.74}$
Effective oil perm. after aging		1	0.5	1.48	1.34	6.88
		2	1	2.78	2.65	7.10
		3	1.5	4.22	4.09	6.95
		linear coef.	0.9992	offset	0.08	$\bar{k} = \mathbf{6.97}$

USS coreflooding expierments DSW injection	
Inj. rate [cm^3/min]	Sw
0.05	0.538
0.2	0.622
1	0.670
3	0.682
5	0.694

Table A.17: USS coreflooding results In9b.

Appendix B

Numerical study appendix

B.1 Source code excerpts - Centrifuge

Figure B.1, Dirichlet boundary implementation: As evaluated in Chapter 8, the centrifuge model uses Dirichlet boundary conditions to define the fluid inflow and outflow. In regard to the primary variables (P_w and So), the implemented model requires a definition of the water phase pressure and oil phase saturation.

Initially, Equation 8.22 calculates the average centrifugal acceleration $g_{c,rm}$ along the core sample. The inlet pressure P_{inlet} is then calculated according to Equation 8.23 (cf. Figure B.1).

The calculation of the inlet pressure P_{inlet} is divided into two parts. In the first place, the build-up time buffers the centrifuge acceleration increase in case the centrifuge spin velocity is changed (cf. Equation 8.36). After a build-up time of 100 *seconds*, the centrifugal acceleration changes to a constant value (Equation 8.22).

In line with Equation 8.24, the Dirichlet outlet boundary condition is defined by a constant pressure and saturation value. The outlet face water saturation remains at a constant value of Sw_{spon} while the outlet (oil) pressure is defined by the atmospheric pressure. The Dirichlet inlet and outlet boundary implementation are summarized in Figure B.1.

B.1 SOURCE CODE EXCERPTS - CENTRIFUGE

Figure B.2, centrifugal acceleration: The centrifugal acceleration is the driving force during the centrifuge experiments and consequently requires an accurate numerical implementation. Figure B.2 summarizes the centrifugal acceleration implementation, which is based on Equation 8.35. While Equation g_c defines the gravity calculation, the DuMu^x implementation requires a comprehensive centrifugal forces calculation for each grid note. The implemented method considers the dynamic centrifugal forces by invoking the global position of each grid cell. The calculated force is then stored inside a mutable gravity vector, which is updated at each time step. In line with the Dirichlet boundary definition, the centrifugal force calculation considers the build-up time during the centrifuge acceleration.

Figure B.3, Hydraulic properties implementation: The benefits of the imbibition adapted Modified capillary pressure model (Equation 8.28) and Modified relative permeability model (Equation 8.30 and Equation 8.31) are discussed in detail in Chapter 8. Since the Modified-P_c parameters B, C, D are defined by the simulation input file, the defined parameters are initially invoked by the P_c method (cf. Figure B.3). The arising P_c is then calculated as function of the normalized/effective water saturation.

Similar to the Modified-P_c law, the Modified-kr law method initially invokes the input parameters ϵ, G, H as well as the endpoint oil and water relative permeabilities. The kr_w and kr_o values are calculated based on Equation 8.30 and Equation 8.31. The implementation of the Modified-P_c and Modified-kr law is summarized in Figure B.3.

B.1 SOURCE CODE EXCERPTS - CENTRIFUGE

```
                    void dirichletAtPos(PrimaryVariables &values,
                    const GlobalPosition &globalPos) const
{
 Scalar t= this->timeManager().time();
//######RPM1#############################
if ((isInletBoundary_(globalPos)) && (t<=BuildUp))
{
Scalar gc_rm=pow((RPM1)*(((t/BuildUp)*2*M_PI)/60),2)*(Centrifuge_Arm-
(Core_length/2));
Scalar Pressure_inlet =Pressure_atm+density_brine*gc_rm*Core_length;

values[Indices::pressureIdx] = Pressure_inlet;  //Definition of water pressure
values[Indices::nPhaseIdx] = (1-IM_sw_i);}      //Definition of oil saturation

else  if ((isInletBoundary_(globalPos)) && (t<=RPM1_time))
{
Scalar gc_rm = pow(((RPM1*2*M_PI)/60),2)*(Centrifuge_Arm-(Core_length/2));
Scalar Pressure_inlet =Pressure_atm+density_brine*gc_rm*Core_length;

values[Indices::pressureIdx] = Pressure_inlet;  //Definition of water pressure
values[Indices::nPhaseIdx] = (1-IM_sw_i);       //Definition of oil saturation
}
//######RPM2#############################
else  if ((isInletBoundary_(globalPos)) && (t<=RPM1_time+BuildUp))
{
Scalar gc_rm=pow((((RPM1+(RPM2-RPM1)*((t_RPM1_time)/BuildUp))*2*M_PI)/60),2)
*(Centrifuge_Arm-(Core_length/2));
Scalar Pressure_inlet =Pressure_atm+density_brine*gc_rm*Core_length;

values[Indices::pressureIdx] = Pressure_inlet;  //Definition of water pressure
values[Indices::nPhaseIdx] = (1-IM_sw_i);       //Definition of oil saturation
}
else  if ((isInletBoundary_(globalPos)) && (t<=RPM2_time))
{
Scalar gc_rm = pow(((RPM2*2*M_PI)/60),2)*(Centrifuge_Arm-((Core_length)/2));
Scalar Pressure_inlet =Pressure_atm+density_brine*gc_rm*Core_length;

values[Indices::pressureIdx] = Pressure_inlet;  //Definition of water pressure
values[Indices::nPhaseIdx] = (1-IM_sw_i);       //Definition of oil saturation
}
//... to be continued until RPM9
//###################################
//Implementation of the outlet boundary, constant during the entire
simulation
if (isOutletBoundary_(globalPos) && (t<=RPM9_time))
{
values[Indices::pressureIdx] = Pressure_atm;  //Definition of water pressure
values[Indices::nPhaseIdx] = (1-OM_sw_i);     //Definition of oil saturation
}
}
```

Figure B.1: Dirichlet boundary implementation.

B.1 SOURCE CODE EXCERPTS - CENTRIFUGE

```cpp
          const GravityVector &gravityAtPos
                  (const GlobalPosition &globalPos) const
{
Scalar t= this->timeManager().time();
Scalar VZ=1; // +1 for imbibition, -1 for drainage

//RPM1
if (t <= (BuildUp))
{g_c[0]=VZ*(pow((((RPM1)*((t/BuildUp))*2*M_PI)/60),2)*(Centrifuge_Arm -
Core_length + (this->bBoxMin()[0] + globalPos[0]))); }
else if (t <= RPM1_time)
{gc_[0]=VZ*(pow(((RPM1*2*M_PI)/60),2)*(Centrifuge_Arm-Core_length+(this-
>bBoxMin()[0] + globalPos[0])));}//Calculation of g_c as a function of position

//RPM2
else if (t <= (RPM1_time+BuildUp))
{g_c[0]=VZ*(pow((((RPM1+(RPM2-RPM1)*((t-RPM1_time)/BuildUp))*2*M_PI)/60),2)
*(Centrifuge_Arm-Core_length+(this->bBoxMin()[0] + globalPos[0]))); }
else if (t <= RPM2_time)
{g_c[0]=VZ*(pow(((RPM2*2*M_PI)/60),2)*(Centrifuge_Arm-Core_length+(this-
>bBoxMin()[0] + globalPos[0]))); }//Calculation of g_c as a function of position

//RPM3
else if (t <= (RPM2_time+BuildUp))
{g_c[0]=VZ*(pow((((RPM2+(RPM3-RPM2)*((t-RPM2_time)/BuildUp))*2*M_PI)/60),2)
*(Centrifuge_Arm-Core_length+(this->bBoxMin()[0] + globalPos[0]))); }
else if (t <= RPM3_time)
{g_c[0]=VZ*(pow(((RPM3*2*M_PI)/60),2)*(Centrifuge_Arm-Core_length+(this-
>bBoxMin()[0] + globalPos[0]))); }//Calculation of g_c as a function of position

//RPM4
else if (t <= (RPM3_time+BuildUp))
{g_c[0]=VZ*(pow((((RPM3+(RPM4-RPM3)*((t-RPM3_time)/BuildUp))*2*M_PI)/60),2)
*(Centrifuge_Arm-Core_length+(this->bBoxMin()[0] + globalPos[0]))); }
else if (t <= RPM4_time)
{g_c[0]=VZ*(pow(((RPM4*2*M_PI)/60),2)*(Centrifuge_Arm-Core_length+(this-
>bBoxMin()[0] + globalPos[0]))); }//Calculation of g_c as a function of position

//… to be continued until RPM9
return g_c;
}

mutable GravityVector  g_c;
```

Figure B.2: Centrifugal acceleration implementation.

B.1 SOURCE CODE EXCERPTS - CENTRIFUGE

```
static Scalar pc(const Params &params, Scalar swe, Scalar salinityPC)
{
        using std::pow;
        using std::min;
        using std::max;
        swe = min(max(swe, 0.0), 1.0);

        Scalar B=  params.B(); //invoking the input paramater
        Scalar C=  params.C(); //invoking the input paramater
        Scalar D=  params.D(); //invoking the input paramater

    return ((B+C*swe)/(1+D*swe)); //calculation and return of Pc values
}
static Scalar krw(const Params &params, Scalar swe, Scalar salinityKRW)
{
        using std::pow;
        using std::min;
        using std::max;
        swe = min(max(swe, 0.0), 1.0);

    Scalar alpha_krw = (params.epsilon_krw()); //invoking the input paramater
    Scalar H_krw     = (params.h_krw());       //invoking the input paramater
    Scalar V_krw     = (params.g_krw());       //invoking the input paramater

    Scalar b_krw     = (2*alpha_krw*(1-H_krw))-V_krw-H_krw;
    Scalar a_krw     = V_krw-H_krw-b_krw;

    Scalar krw_max   = params.krw_max(); //invoking the input paramater
return
(krw_max*(((a_krw/(2*alpha_krw))*(pow(swe,(2*alpha_krw)))+(((b_krw/alpha_krw)
*pow(swe,(alpha_krw)))*pow(swe,(H_krw)))))); //calculation and returning krw
}
static Scalar krn(const Params &params, Scalar swe,Scalar salinityKRO)
{
        using std::pow;
        using std::min;
        using std::max;
        swe= min(max(swe, 0.0), 1.0);

    Scalar alpha_kro = (params.epsilon_kro()); //invoking the input paramater
    Scalar H_kro     = params.h_kro();         //invoking the input paramater
    Scalar V_kro     = (params.g_kro());       //invoking the input paramater

    Scalar b_kro = (2*alpha_kro*(1-H_kro))-V_kro-H_kro;
    Scalar a_kro = V_kro-H_kro-b_kro;

    Scalar kro_max = params.kro_max();    //invoking the input paramater
    Scalar swe_o=(1-swe);

return
(kro_max*((a_kro/(2*alpha_kro))*(pow(swe_o,(2*alpha_kro)))+(((b_kro/alpha_kro
)*pow(swe_o,(alpha_kro)))*pow(swe_o,(H_kro)))));//calculation and return}
```

Figure B.3: Modified-P_c and Modified-kr law implementation.

B.2 Source code excerpts - Coreflooding

Figure B.4, Neumann inlet boundary implementation: While Dirichlet boundary conditions require a pressure and saturation specification, Neumann boundaries are defined by source or sink terms.

The Neumann inlet boundary formulation of the coreflooding model is defined by the experimentally applied injection rates. However, the applied volumetric rates of $0.05, 0.2, 1, 3$ and $5\,cm^3/min$ require the adaption to the numerical 2D model. The required unit of $[kmol/(s \cdot m^3)]$ is obtained by inserting the volumetric brine injection rates into Equation 9.1 (cf. Figure B.4). After converting the injection rate into 2D equivalent, the injected water mass q_w^w and the corresponding salt concentration q_w^{NaCl} are specified. The mass injection is adapted to the corresponding time duration and laboratory injection rate.

Figure B.5, Neumann outlet boundary implementation: The outlet boundary formulation represents the key challenge of the coreflooding simulation. A simplification and adaption of Peaceman's well model revealed to be representative of the mass outflow. The quantity of the sink term is thereby mainly effected by the occurring phase mobilities and the experimentally set production pressure P_{Prod}. Equation 9.4 specifies the quantity of q_w^w and Equation 9.5 specifies the quantity of q_o^o. The salt component production q_o^{NaCl} is in balance to the water phase extraction. The DuMu^x implementation is displayed in Figure B.5.

B.2 SOURCE CODE EXCERPTS - COREFLOODING

```cpp
void solDependentNeumann(PrimaryVariables &values,
            const Element &element,
            const FVElementGeometry &fvGeometry,
            const Intersection &intersection,
            const int scvIdx,
            const int boundaryFaceIdx,
            const ElementVolumeVariables &elemVolVars) const
{
    GlobalPosition globalPos;
    if (isBox)
    globalPos = element.geometry().corner(scvIdx);
    else
    globalPos = intersection.geometry().center();

    Scalar t= this->timeManager().time();
    const FluidState &fs = elemVolVars[scvIdx].fluidState();
    values = 0;

//Inlet boundary implementation

if (isInletBoundary_(globalPos))
{ //Conversion from "3D" lab injection rate into "2D" injection rate
Scalar qw_w=-(((qw_lab/60)*1e-6)*(fs.density(wPhaseIdx))*
(1/FluidSystem::molarMass(wCompIdx)))/(M_PI*(pow((Core_diameter/2),2)));

if (t<=inj_time_005)
 {values[contiH2OEqIdx]   = 1*qw_mol; //Definition of qw_w mass injection
 //admixture of qw_nacl concentration based on injection brine composition
 values[contiNaClEqIdx] = 1*qw_mol*(massTomoleFrac_(SeaWaterSalinity));
 }

else if (t<=inj_time_02)
{
values[contiH2OEqIdx]= 4*qw_mol; //Definition of qw_w mass injection
//admixture of qw_nacl concentration based on injection brine composition
values[contiNaClEqIdx] =  4*qw_mol*(massTomoleFrac_(SeaWaterSalinity)); }

else if (t<=inj_time_1)
{
values[contiH2OEqIdx]= 20*qw_mol; //Definition of qw_w mass injection
//admixture of qw_nacl concentration based on injection brine composition
values[contiNaClEqIdx] = 20*qw_mol*(massTomoleFrac_(SeaWaterSalinity)); }

// to be continued until 5 cc/min
 }
```

Figure B.4: Neumann inlet boundary implementation.

B.2 SOURCE CODE EXCERPTS - COREFLOODING

```cpp
void solDependentNeumann(PrimaryVariables &values,
            const Element &element,
            const FVElementGeometry &fvGeometry,
            const Intersection &intersection,
            const int scvIdx,
            const int boundaryFaceIdx,
            const ElementVolumeVariables &elemVolVars) const
    {
        GlobalPosition globalPos;
        if (isBox)
        globalPos = element.geometry().corner(scvIdx);
        else
        globalPos = intersection.geometry().center();

        Scalar t= this->timeManager().time();
        const FluidState &fs = elemVolVars[scvIdx].fluidState();
        values = 0;
// … inlet boundary formulation…
//Outlet formulation is constant during the entire simulation
else if (isOutletBoundary_(globalPos) && (t<=inj_time_5))
{

//Calculation of qw_w_outlet
Scalar qw_w_outlet=(((PermMultiplicatorOutlet*Permeability*9.8692338*1e-16)
*(sqrt(fvGeometry.subContVol[scvIdx].volume))*((fs.molarDensity(wPhaseIdx))
*(elemVolVars[scvIdx].mobility(wPhaseIdx))))
*(fs.pressure(wPhaseIdx)-ProductionPressure_)*
(1/fvGeometry.subContVol[scvIdx].volume));

//Calculation of qo_o_outlet
Scalar qo_o_outlet=(((PermMultiplicatorOutlet*Permeability*9.8692338*1e-16)
*(sqrt(fvGeometry.subContVol[scvIdx].volume))*((fs.molarDensity(nPhaseIdx))
*(elemVolVars[scvIdx].mobility(nPhaseIdx))))
*(fs.pressure(nPhaseIdx)-ProductionPressure_)*
(1/fvGeometry.subContVol[scvIdx].volume));

values[contiH2OEqIdx]=qw_w_outlet;    Definition of qw_w_outlet
values[contiOILWOSEqIdx]=qo_o_outlet; Definition of qo_o_outlet


//Definition of qw_nacl_outlet
//Extraction of salt component in balance to qw_w
values[contiNaClEqIdx] =qo_o_outlet*(fs.moleFraction(wPhaseIdx,NaClIdx));
}
}
```

Figure B.5: Neumann outlet boundary implementation.

Anhang C

Ausführliche Zusammenfassung

C.1 Einleitung

Im Jahr 2013 wurde eine Kooperation zwischen dem Petroleum Institute Abu Dhabi und dem Institut für Erdöl und Erdgastechnik Clausthal geschlossen. Im Rahmen dieser Zusammenarbeit wurde eine umfassende experimentelle und numerische Low-salinity Studie in Kalksteinproben realisiert, die in die Verfassung der hier vorgelegten Dissertation resultierte. Der überwiegende Teil der Dissertation wurde während eines dreieinhalbjährigen Forschungsaufenthaltes am Petroleum Institute Abu Dhabi angefertigt.

Während die Studie dabei von den hervorragenden experimentellen Forschungsmöglichkeiten am Petroleum Institute profitierte, ist das Wissen des Instituts für Erdöl und Erdgastechnik Clausthal vor allem in die Umsetzung der numerischen DuMuX Simulationen eingeflossen.

Der World Energy Council [40] prognostiziert einen Anstieg des Primärenergieverbrauches um 27 bis 61 % bis zum Jahr 2050. Obwohl die Entwicklung und Umsetzung erneuerbarer Energiesysteme den Energieerzeugungsmix erheblich verändern, bleibt die Abhängigkeit von den fossilen Brennstoffen Kohle, Öl und Gas weiterhin hoch. Es

C.2 LITERATURRECHERCHE

wird erwartet, dass im Jahr 2050 noch rund ein Viertel des Primärenergieverbrauchs durch die Verbrennung von Kohlenwasserstoffen gedeckt wird.

Da allgemein angenommen wird, dass der Großteil der vorhandenen Ölreserven entdeckt wurde, konzentriert sich die Ölindustrie zunehmend auf die Verbesserung der Lagerstättenausbeutung. Als Alternative zu den teureren chemischen Enhanced Oil Recovery Methoden wie Polymer- oder Tensidfluten, untersucht diese Arbeit das Potenzial von Low-salinity Wasserfluten um die Ölgewinnung zu erhöhen.

Das Konzept des Low-salinity Wasserfluten zielt dabei auf die Änderung der Lagerstättengesteinsbenetzbarkeit ab. Während Kalksteinlagerstätten im Allgemeinen ölbenetzende Eigenschaften aufzeigen, kann die signifikante Reduzierung des Lagerstättensalzgehaltes die Gesteinsbenetzbarkeit in Richtung stärker wasserbenetzten Bedingungen verändern.

Obwohl Low-salinity Wasserfluten im Allgemein einen positiven Einfluss auf die Ölgewinnung zugeschrieben wird, sind die involvierten physikalischen und chemischen Mechanismen umstritten. Um das Verständnis von Low-salinity Wasserfluten zu verbessern, wurde im Rahmen dieser Dissertation eine umfassende experimentelle und numerische Studie umgesetzt.

Die experimentellen Ergebnisse dieser Dissertation wurden auf dem Symposium der Society of Core Analysis in Trondheim, Norwegen, vorgestellt [34]. Eine Beschreibung der mathematischen und numerischen Modelle ist im Journal of Petroleum Science and Engineering [35] veröffentlicht.

C.2 Literaturrecherche

Das Kapitel Literaturrecherche stellt zunächst das Low-salinity Konzept am Beispiel der analytischen Buckley-Leverett Lösung vor. Neben dem konventionellen Wasserfluten als Referenzbeispiel, werden die sekundäre und tertiäre Buckley-Leverett Low-salinity Lösung vorgestellt.

C.2 LITERATURRECHERCHE

Die Zusammenfassung der derzeitigen Low-salinity Forschung in Limestones ist in drei Unterkapitel aufgeteilt: Spontaneous-imbibition Tests, Kernflutungsexperimente sowie Low-salinity Mechanismen. In Übereinstimmung mit dem Titel der Dissertation konzentriert sich die Literaturrecherche dabei ausschließlich auf Low-salinity Effekte in Limestone Proben.

Die Low-salinity Forschung verfolgt derzeit zwei Ansätze um die Ölproduktion in Kalksteinen zur verbessern. Erstens wird angenommen, dass die Modifikation der Lagerstättensalinität (im Besonderen die Konzentration von Potentialbestimmende Ionen, PDI) zur Abtrennung von Öl von der positiv geladenen Kalksteinoberfläche führen kann. Zweitens wurde gezeigt, dass die Injektion von stark entsalztem Wasser die Ölgewinnung verbessern kann [89].

Die Zusammenfassung der veröffentlichen Spontaneous-imbibition Tests in Limestones deutet an, dass entsalzte Wasserlösungen ein enormes Potential aufweisen, um spontane Ölproduktion hervorzurufen. Im Vergleich zu PDI angereichtes Injektionswasser, deutete stark entsalztes Injektionswasser ein signifikant höheres sekundäres und tertiäres Ölgewinnungspotenzial an [5, 61, 82, 89, 107].

Neben Spontaneous-imbibition Tests beinhaltet die Fachliteratur mehrere Beispiele von Low-salinity Kernflutungsexperimenten. Während Al-Attar et al. [4], Shehata et al. [91] und Tetteh et al. [99] eine verbesserte Ölgewinnung durch sulfatreiche Injektionswasser beobachten, berichtet die Mehrzahl der veröffentlichen Studien [22, 45, 82, 106, 108, 109], dass die Verringerung der Ionenstärke der vielversprechende Ansatz zur Ölgewinnung in Kernflutungseperimenten ist.

Die Forschungsgruppe von Mahani et al. [14, 64, 65, 66] schlägt Surface-charge-change als den treibenden Mechanismus von Low-salinity Effekten in Kalksteinen vor. In Abhängigkeit der Salzzusammensetzung nimmt die Kalksteinoberfläche eine positive oder negative Oberflächenladung ein. Während eine hohe Salinität in eine positive Oberflächenladung resultiert (und damit in eine Bindung von Öl), bewirkt ein salzfreies System in eine negative Oberflächenladung (und damit die Abtrennung von Öl).

C.3 Probenvorbereitung

Eine sorgfältige und einheitliche Probenvorbereitung ist entscheidend für den Erfolg einer experimentellen Studie. In diesem Kapitel wird der Umgang und die Vorbereitung der verwendeten Fluide, die Behandlung der Kernproben und die Probeninitialisierung beschrieben.

Die Öl- und Salzwasserdichten (Formationswasser, Meerwasser und verdünntes Meerwasser) sowie die Öl- und Salzwasserviskositäten wurden in einem Temperaturbereich von 20 bis 50°C bzw. 30 bis 60°C gemessen. Während eine lineare Gleichung die Fluiddichten auf 70°C interpoliert, wurde die Arrhenius-Gleichung verwendet, um die Fluidviskositäten auf 70°C zu interpolieren.

Die Grenzflächenspannung zwischen Formationswasser und Öl, Meerwasser und Öl, sowie verdünntem Meerwasser und Öl wurden mit Hilfe der Pendant-Drop Methode gemessen. Die Unterschiede der Grenzflächenspannung zwischen dem verschiedenen Salzwasser-Öl Systemen sind gering.

Die Studie verwendet Indiana Limestone Proben um Spontaneousimbibition Tests, Zentrifugen und- Kernflutungseperimenten durchzuführen. Auf Grund der Zentrifugenabmaße ist die Probenlänge auf $5\,cm$ begrenzt. Calcit stellt die Hauptkomponente der Indiana Limestone Proben dar (98,6 $wt\%$).

Die Kerne wurden mit Hilfe eines Pore-flush-cleaning System gereinigt. Nachdem die Kerne gesättigt und Sw_c erreicht wurde, wurden die Kerne 30 Tage lang gealtert. Die gemessen Sw_c Werte liegen zwischen 23,0 und 32,5 %, was mit den NMR-basierten Referenzwerten übereinstimmt. Da die effektive Permeabilität vor und nach dem Altern abgenommen hat, ist anzunehmen, dass die Kerne zu Anfang der Studie ölbenetzende Eigenschaften aufwiesen.

Ein simples und einheitliches Permeameter wurde entwickelt um die absolute und effektive Permeabilität zu messen. Alle Messungen zeigten eine lineare Korrelation zwischen den angewendeten Einspritzraten und den entsprechenden Differenzdrücken. Die Ergeb-

C.4 SPONTANEOUS-IMBIBITION TESTS

nisse der absoluten und effektiven Permeabilitätsmessungen sind in Anhang A aufgeführt.

Nach Abschluss der Spontanen- und Forced-Imbibition Experimente wurden die Kalksteinproben gereinigt, gesättigt und für die nachfolgenden Experimente initialisiert. Die Reproduktion der anfänglichen absoluten und effektiven Permeabilitätswerte deuten auf ein verlässliches Kernvorbereitungsablauf hin.

C.4 Spontaneous-imbibition Tests

Im Rahmen der Low-salinity Studie wurden insgesamt zwölf Spontaneous-imbibition Tests durchgeführt. Neben dem Einfluss des Injektionswassers wurde auch der Einfluss des Sedimentwassers auf die spontane Ölgewinnung untersucht. Die verschiedenen Sedimentwasser und Injektionswasser Kombinationen wurden in vier Gruppen unterteilt: Gruppe I, Identischer Salzgehalt von Sedimentwasser und Injektionswasser, Gruppe II und III, stark salzhaltiges Sedimentwasser und niedrig salzhaltiges Injektionswasser und Gruppe IV, niedrig salzhaltiges Sedimentwasser und stark salzhaltiges Injektionswasser.

Sofern das Sedimentwasser und das Injektionswasser die gleiche Zusammensetzung hatten (Gruppe I), wurde eine marginale spontane Ölgewinnung von 1,5 % für Lagerstättenwasser, 2,3 % spontane Ölgewinnung für Meerwasser und 2,9 % spontane Ölgewinnung für verdünntes Meerwasser beobachtet. Die Kombination eines stark salzhaltigen Sedimentwasser und eines niedrig salzhaltigen Injektionswasser (Gruppe II und Gruppe III) führte zu signifikanter spontaner Ölgewinnung. Lagerstättenwasser (CW) und Meerwasser (IW) verursachten eine spontane Ölgewinnung von 22,1 %, Lagerstättenwasser (CW) und verdünntes Meerwasser (IW) verursachten eine spontane Ölgewinnung von 35,4 % und Meerwasser (CW) und verdünntes Meerwasser (IW) führten zu einer spontanen Ölgewinnung von 22,4 %. Die Kombination eines niedrig salzhaltiges Sedimentwasser und eines stark salzhaltiges Injektionswasser verursachte keine spontane Ölgewinnung (Gruppe IV).

C.5 ZENTRIFUGENVERSUCHE

Es kann daher geschlussfolgert werden, dass spontane Ölgewinnung nur bei der Kombination eines stark salzhaltigen Sedimentwasser und einem niedrig salzhaltigem Injektionswasser vorkommt. Keine oder kaum spontane Ölproduktion wurde beobachtet, sofern das Injektionswasser einen höheren oder den gleichen Salzgehalt wie das Sedimentwasser hatte. Die Ergebnisse der Spontaneous-imbibition Tests sind in Tabelle 5.2 und Figure 5.5 bis Figure 5.5 zusammengefasst.

In Übereinstimmung mit den Spontaneous-imbibition Tests zeigten die Kontaktwinkelmessungen der verschiedenen Systeme eine Änderung der Benetzbarkeit in Richtung stärkerer wasserbenetzende Eigenschaften an, wenn Meerwasser oder verdünntes Meerwasser im Öl-Wasser-Feststoff-System vorhanden waren.

Die Arbeit von Zhang et al. [111] und Mahani et al. [64, 65] schlägt eine Änderung der Oberflächenladung als möglichen Mechanismus von Low-salinity Effekten in Kalksteinen vor. Obwohl der verwendete Zeta-potential Messaufbau keine Potentialmessungen von unverdünnten Formations- und Meerwassersystemen ermöglichte, zeigten die schrittweise Salzwasserverdünnung und die entsprechenden Zeta-potential Werte eine deutliche Tendenz. Die Verringerung der Salinität bewirkt eine Änderung der Kalksteinoberflächenladung in Richtung stärkerer negativer Werte. In Übereinstimmung mit der Arbeit von Mahani et al. [64, 65] kann angenommen werden, dass eine stärkere negative Oberflächenladungen die Ablösung des Öls von der Kalksteinoberfläche verursacht.

C.5 Zentrifugenversuche

Nachdem das Spontaneous-imbibition Verhalten der verschiedenen Sedimentwasser (CW) und Injektionswasser (IW) Kombinationen getestet wurde, wurde der Einfluss von Salinität auf das Forced-Imbibition Verhalten in Kalksteinproben untersucht.

Zunächst stellt das Kapitel die experimentelle Durchführung der Zentrifugenversuche dar. Dabei wurde die analytische Hassler & Brunner

C.6 KERNFLUTUNGSEXPERIMENTE

Korrektur hergeleitet, um die Notwendigkeit einer zuverlässigen Durchschnitt- zu Einflusswassersättigung Korrektur zu veranschaulichen. Zudem beschreibt die Arbeit die erfolgreiche Kombination einer hyperbolischen Funktion und Forbes-first-solution um Imbibition Daten effizient und einheitlich auszuwerten.

Die Imbibition Kapillardruckkurven zeichnen sich durch eine zunehmende Wasserbenetzung und eine gleichzeitige Verringerung der Restölsättigung aus, sobald der Salzgehalt der gemessenen Systeme abnimmt. Unter Verwendung von Formationswasser als Sedimentwasser resultierten Formationswasser (IW), Meerwasser (IW) und verdünntes Meerwasser (IW) in eine Forbes-first-solution korrigierte Restölsättigung von 15,3 %, 10,7 % und 5,6 %. Die höchste Ölgewinnung wurde beobachtet, wenn verdünntes Meerwasser als Sedimentwasser und Injektionswasser verwendet wurde (Restölsättigung von 3,2 %). Die Testergebnisse sind in Tabelle 6.2 und in Figure 6.4 bis Figure 6.4 zusammengefasst.

C.6 Kernflutungsexperimente

Die durchgeführten Spontaneous-Imbibition und Forced-Imbibition Experimente verdeutlichen eine Verbindung zwischen Salzgehalt, Benetzbarkeit, Kapillardruck sowie Restölsättigung. Um den Einfluss der Salinität auf die Ölgewinnung unter möglichst realistischen Lagerstättenbedingungen zu evaluieren, wurden Formationswasser, Meerwasser und verdünntes Meerwasser zudem in Kernflutungsxperimenten getestet.

Zu Beginn des Kapitels werden das physikalische und experimentelle Konzept von (unsteady state) Kernflutungsexperimenten beschrieben. Die verschiedenen Salzwasserzusammensetzungen wurden zunächst mit einer Injektionsrate von $0.05\, cm^3/min$ injiziert was einer typischen Lagerstättenflussgeschwindigkeit von $1\, feet/day$ entspricht. Die daraufhin folgenden Injektionsraten von $0.2, 1, 3$ und $5\, cm^3/min$ schwächen sukzessive den Einfluss des Capillary-end-effects auf die Ölförderung ab.

C.7 NUMERISCHE ZENTRIFUGENSIMULATION

Insgesamt wurden drei Kernflutungsexperimente durchgeführt. Die sekundäre Injektion von Formationswasser (1), die sekundäre Injektion von Meerwasser (2) und die sekundäre Injektion von verdünntem Meerwasser (3). Abhängig vom Versuchsaufbau wurden 90 bis 180 Porevolumen bei einer Temperatur von 70°C injiziert. Bei einer sekundären Injektionsrate von 0.05 cm^3/min verursachte Formationswasser 21,9 %, Meerwasser 30,1 % und verdünntes Meerwasser verursachte 38,5 % Ölgewinnung. Die endgültige sekundäre Restölsättigung betrug 37,4 % für Formationswasser, 35,5 % für Meerwasser und 30,6 % für verdünntes Meerwasser. Eine geringfügige zusätzliche Ölrückgewinnung trat in einem von zwei Kernflutungsexperimenten auf, sofern verdünntes Meerwasser in tertiären Injektionsmodus injiziert wurde. Die Ergebnisse der Kernflutungsexperimente sind in Tabelle 7.1 und in Figure 7.4 bis Figure 7.6 zusammengefasst.

C.7 Numerische Zentrifugensimulation

Die Besonderheiten der Zentrifugenmethode erfordert eine analytische und vorzugsweise numerische Auswertung der experimentell erhaltenen Daten. Spezielle SCAL-Software wie Cydar oder Sendra bieten numerische Möglichkeiten, mit denen die Zentrifugenmethode simuliert werden kann [60].

Um jedoch ein vollständiges Verständnis der numerischen Simulationen zu gewährleisten, entwickelt diese Arbeit ein unabhängiges numerisches Zentrifugen- und Kernflutungsmodell. Dabei wurde der Open-source C++ Simulator DuMu^x als Grundlage benutzt. Die entwickelten numerischen Zentrifugen- und Kernflutungsmodelle wurden mit Hilfe der kommerziellen Software Cydar verifiziert.

Dabei definieren Dirichlet-Randbedingungen den Fluidstrom an der Einstrom- und Ausstromfläche

$$P_{inlet} = P_{atm} + \rho_w \cdot g_{c,rm} \cdot L, \quad where \ Sw_{inlet} = 1, \quad (C.1)$$

und

C.8 NUMERISCHE KERNFLUTUNGSSIMULATION

$$P_{outlet} = P_{atm}, \quad \text{where} \ Sw_{outlet} = Sw_{spon}. \quad (C.2)$$

Die Massenbilanz der Zentrifugenmodells kann wie folgt zusammengefasst werden

$$\phi \frac{\partial \left(\rho_o S_o + \rho_w S_w \right)}{\partial t} + \nabla \cdot (\rho_w \, v_w + \rho_o v_o) = q_i, \quad i = o, w, \quad (C.3)$$

wobei die Zentrifugalkraft über die advektive Geschwindigkeit berücksichtigt ist

$$v_i = -\frac{k \cdot kr_i}{\mu_i} \cdot (\nabla P_i - \rho_i g_c), \quad i = o, w, \quad (C.4)$$

und

$$g_c(r) = \omega^2 \cdot r = \left(\frac{RPM \cdot 2\pi}{60} \right)^2 \cdot r. \quad (C.5)$$

Das entwickelte Modell wurde erfolgreich gegen die kommerzielle Cydar-Software validiert. Die numerischen Simulationsergebnisse bestätigen den analytisch erhaltenen Kapillardruck und die experimentell erhaltene Restölsättigung. Der Einfluss der relativen Permeabilität auf die History-matches ist auf die Steigung der Sättigungskurve begrenzt. Während einige Studien die Zentrifugenmethode verwenden, um relative Permeabilitätskurven numerisch abzuleiten, führt diese Studie Kernflutungsexperimente durch, um relative Permeabilitätsdaten zu bestimmen. Die Ergebnisse der numerischen Simulationen sind in Figure 8.4 bis Figure 8.10 zusammengefasst.

C.8 Numerische Kernflutungssimulation

Während die Zentrifugenexperimente mit konstanten Fluideigenschaften simuliert wurden, beinhalteten die Kernflutungsexperimente In2b

C.8 NUMERISCHE KERNFLUTUNGSSIMULATION

und In4b den Austausch des Injektionswasser. Dabei beeinflusst die sich veränderte Viskosität erheblich den auftretenden Differenzialdruck. Um die Auswirkungen der dynamischen Fluideigenschaften zu berücksichtigen, wurde eine 2-Phasen-3-Komponenten Fluidmodell umgesetzt. Die Ölphase besteht dabei aus einer einzigen Ölkomponente, während die Wasserphase einen Wasserkomponenten und einen (Pseudo-) Salzkomponenten enthält. Die experimentell gemessene Viskosität und Dichte von Formationswasser, Meerwasser und verdünntem Meerwasser werden dabei als Funktion der Salzkomponente interpoliert.

Die Neumann Randbedingungen für die Kernflutgungsexpiermente sind im folgenden zusammengefasst

$$q_w^w = \frac{k}{V_G} \cdot \frac{\hat{\rho}_w kr_w}{\mu_w} \cdot (P_w - P_{Prod}), \tag{C.6}$$

und für den Salzkomponenten

$$q_w^{NaCl} = q_w^w \cdot x_w^{NaCl}. \tag{C.7}$$

Während die Menge der injizierten Masse an der Einflussfläche durch die angewendeten Injektionsraten definiert ist, ist die mathematische Formulierung der Ausflussrandbedingung wesentlich komplexer. Diese Arbeit nutzt eine Neumann-Randbedingung, die die Massenentnahme in Abhängigkeit der Phasenmobilität und dem experimentel definiertem Produktionsdruck berechnet

$$q_w^w = \frac{k}{V_G} \cdot \frac{\hat{\rho}_w kr_w}{\mu_w} \cdot (P_w - P_{Prod}), \tag{C.8}$$

$$q_w^{NaCl} = \underbrace{\frac{k}{V_G} \cdot \frac{\rho_w kr_w}{\mu_w} \cdot (P_w - P_{Prod})}_{q_w^w} \cdot x_w^{NaCl}, \tag{C.9}$$

C.9 ZUSAMMENFASSUNG

und für die Ölkomponente

$$q_o^o = \frac{k}{V_G} \cdot \frac{\hat{\rho}_o k r_o}{\mu_o} \cdot (P_o - P_{Prod}). \quad \text{(C.10)}$$

Die Massenbilanz der Kernflutungen kann wie folgt zusammengefasst werden

$$\phi \frac{\partial \left(\hat{\rho}_o x_o^k S_o + \hat{\rho}_w x_w^k S_w \right)}{\partial t} + \nabla \cdot \left(\hat{\rho}_w x_w^k v_w + J_w^k + \hat{\rho}_o x_o^k v_o \right) = q_i^k \cdot \text{(C.11)}$$
$$i = o, w, \quad k = o, w, NaCl.$$

Zunächst wurde das numerische Kernflutungsmodell anhand der absoluten Permeabilitätsmessungen verifiziert (Table 9.1). Zudem zeigen die implementieren Kernflutung History-matches eine gute Übereinstimmung zwischen der DuMu^x und Cydar Simulation (Figure 9.3 bis Figure 9.10). Die Wassersättigung, der Wasserphasendruck und der Kapillardruck der numerischen Kernflutunggssimulation In4b verdeutlichen die Einhaltung der auferlegten Randbedingungen (Figure 9.7 bis Figure 9.9). Darüber hinaus bestätigen die History-matches die physikalische Plausibilität der experimentell durchgeführten Kernflutungen. Im sekundären Injektionsmodus zeigen die numerisch erhaltenen relativen Permeabilitäts- und Kapillardruckkurven das stärkste wasserbenetzende Verhalten für verdünntes Meerwasser, gefolgt von Meerwasser und Formationswasser.

C.9 Zusammenfassung

Die Spontaneous-imbibition-, Zentrifugen- und Kernflutungsexperimente verdeutlichen eine Korrelation zwischen Salzgehalt, Benetzbarkeit und Ölgewinnung in Kalksteinproben. Im Vergleich zu Formationswasser und Meerwasser, zeigte verdünntes Meerwasser das mit Abstand vielversprechendste Potenzial zur effizienten Ölgewinnung. Unter Verwendung von Formationswasser als Sedimentwasser führte verdünntes Meerwasser zu einer spontanen Ölgewinnung

C.9 ZUSAMMENFASSUNG

von 35,4 %, Meerwasser zu einer spontanen Ölgewinnung von 22,1 % und Formationswasser zu einer spontanen Ölgewinnung von 1,5 %. Darüber hinaus resultierte die Forced-Imbibition in eine Restölsättigung von 5,6 % für verdünntes Meerwasser, 10,7 % für Meerwasser und 15,3 % für Formationswasser. Die Kernflutungsexperimente bestätigten den Einfluss des Injektionswasser auf die Ölgewinnung. Während verdünntes Meerwasser die signifikant höchste Ölgewinnung bei der Injektionsrate von 0.05 cm^3 min verursachte, betrug die verbleibende Restölsättigung (nach dem Rate-Bumping) 30,6 % für verdünntes Meerwasser, 35,5 % für Meerwasser und 37,4 % für Formationswasser.

Die Zeta-potential Messungen stützen die These der Änderung der Oberflächenladung als treibenden Mechanismus von Low-salinity Effekten in Kalksteinen. Im Vergleich zu Formationswasser und Meerwasser, zeigte das System aus verdünntem Meerwasser und Kalkstein das stärkste negative Zeta-potential. In Übereinstimmung mit der Literaturzusammenfassung, den Sponatenoues-imbibition-tests, Zentrifugen- und Kernflutungsexperimenten ist die Änderung der Oberflächenladung daher eine plausible Erklärung für die Änderung der Benetzbarkeit in Kalksteinen.

Die transparente Entwicklung eines numerischen Zentrifugen- und Kernflutungsmodells stellt das umfassende Verständnis der implementierten numerischen Simulationen sicher. Die Modellformulierung umfasst die auferlegten Randbedingungen, Fluideigenschaften, hydraulischen Eigenschaften, Strömungsmodell und Materialbilanzkopplung. Die numerische Simulation verdeutlicht, dass eine einfache mathematische Formulierung ausreicht, um die experimentellen Daten zu validieren.

Die numerische Zentrifugensimulation bestätigt die analytische Zentrifugendatenanalyse. Die Kapillardruckkurven zeigen eine zunehmende Tendenz zur Wasserbenetzung und eine gleichzeitige Verringerung der Restölsättigung, sobald der Salzgehalt des Injektionswasser im Vergleich zu Formationswasser abnimmt. In Übereinstimmung mit den Zentrifugenergebnissen bestätigten die numerischen Kernflutungssimulationen eine Korrelation zwischen Salzgehalt, Be-

C.9 ZUSAMMENFASSUNG

netzbarkeit und Ölgewinnung. Im sekundären Injektionsmodus zeigt die numerisch erhaltene relative Permeabilität das stärkste wasserbenetzende Verhalten für verdünntes Meerwasser, gefolgt von Meerwasser und Formationswasser.